Chinese Agriculture in the 1930s

Hao Hu • Funing Zhong
Calum G. Turvey
Editors

Chinese Agriculture in the 1930s

Investigations into John Lossing Buck's Rediscovered 'Land Utilization in China' Microdata

palgrave
macmillan

Editors
Hao Hu
College of Economics and Management
Nanjing Agricultural University
Nanjing, China

Funing Zhong
College of Economics and Management
Nanjing Agricultural University
Nanjing, China

Calum G. Turvey
Cornell University
Ithaca, NY, USA

ISBN 978-3-030-12687-2 ISBN 978-3-030-12688-9 (eBook)
https://doi.org/10.1007/978-3-030-12688-9

Cover credit © Chronicle / Alamy Stock Photo
Cover designed by Fatima Jamadar

This Palgrave Macmillan imprint is published by the registered company Springer Nature Switzerland AG.
The registered company address is: Gewerbestrasse 11, 6330 Cham, Switzerland

To our wives

Preface

This book is comprised of a number of papers drawn from the works of John Lossing Buck's 1937 publication of *Land Use in China*. Buck's study was based on a survey of 16,786 farms, 168 localities, and 38,256 farm families in 22 provinces in China between 1929 and 1933. It was a landmark study in the sense that no previous widescale study had attempted to provide a geographic and economic map of Chinese agriculture.

Although frequent references have been made to Buck's study, these focused on the three statistical volumes prepared by Buck and his colleagues. That work was prepared by abacus, and the final written volume was driven largely by the statistical information and cross-tabulations that could be taken at the time.

Where does this book fit in? Around the year 2000 a number of paper-wrapped packages were uncovered at the Nanjing Agricultural University (NJAU). The packages contained the actual paper spreadsheets with the individual farm records of some 10,000+ farm households collected by Buck and his team. Immediate efforts were taken by a team of Chinese and Japanese scholars to preserve the records, first by photographic reproduction and then a years-long effort to digitize the data into a useable electronic format. From around 2007, enough data had been recorded to deploy modern economic thought and econometric analysis to start bringing the data to life. It would take nearly another decade to complete the digitization effort, check and verify the data, and prepare it in a form that could be used to empirically, and statistically, evaluate the economics of

Chinese agriculture in the 1930s. This book contains 14 chapters, including the introduction and conclusion. Three chapters are historical in nature. Chapter 2 provides a biopic of Buck, how he arrived in China with a degree in agriculture from Cornell University as an agricultural missionary, landed a faculty position at the University of Nanjing to start a program of study in agricultural economics, started to implement small- and large-scale surveys, including those that led to his two books, *Chinese Farm Economy* in 1930 and *Land Utilization in China* in 1937, and his broader efforts to improve and understand Chinese agriculture. The socioeconomic and political conditions facing Chinese farmers in the Republican era were largely normalized in Buck's studies—taken as given and understood. But in the modern era it is difficult to truly understand and interpret Buck's work without some semblance of understanding of the calamities, catastrophes, and conflicts in the Republican era generally, and the period of study, 1929–1933 in particular. Chapter 3 provides an overview of these conditions. As illustrated by the many footnotes to English language periodicals, this chapter was prepared and compiled from contemporaneous news reports. We had some hesitation about including this chapter in the book, but ultimately decided to err on the side of caution so that the works that follow can be placed in a proper context by the reader.

Chapter 4 may be viewed as a case study of economic archaeology. In this chapter the authors summarize the many years of effort put into compiling and checking the discovered microdata. It details, as accurately as possible, what is believed to have happened to the data records over the years, the circumstances under which they were discovered, the efforts at data recovery, and finally the painstaking efforts at ensuring that once digitized, the data matched the actual summaries provided in Buck's statistical volumes. Incredibly, the efforts confirmed the accuracy of Buck's tables, even to the point of standardizing weights and measures. Importantly, the chapter identifies the specific variables actually recovered. Unfortunately, for reasons unknown, many of the original worksheets were not recovered, so that the chapters that follow faced several data constraints that may affect statistical and econometric reliability.

The remaining chapters represent the analyses done to date. Chapter 5 addresses some of the reliability issues. Noting that some academics questioned Buck's survey techniques, Funing Zhong, Hao Hu, and Qun Su undertook a comparison of Buck's data with other data sources and conclude that the criticisms are not justified, and that Buck's data falls within the ranges of previously documented metrics.

Perhaps one of the more important contributions of this study is the ability to compare a historical representation of Chinese agriculture in the 1930s to the present era. Much, of course, has changed since the Republican era, including the transformative post-revolutionary period between 1949 and 1978, and the reform era that followed.

The remaining chapters are thematically ordered in terms of tenancy, labor and labor efficiency, agricultural production and production efficiency, and credit, all of which were important economic issues facing Chinese agriculture in the 1930s.

On farm tenancy and labor, Chap. 6 by Minjie Yu and Hao Hu examines tenancy issues; Chap. 7 by Hao Hu and Weiwei Zheng evaluates regional differences in surplus agricultural labor; Chap. 8 by Hao Hu and Zhongwei Yang investigates poverty and inequality by providing measures of the Gini coefficient and Engel coefficients.

Production and production efficiency are explored in Chap. 9 where Hisatoshi Hoken and Qun Su examine the data using Box–Cox transformation to determine if there is evidence for an inverse relationship between crop yields and farm size. Chapter 10 by Hao Hu and Minjie Yu tackles the problem of economies of size and scale by generating production coefficients on land and labor to determine elasticities and the substitution between land and labor.

Credit issues are explored in Chap. 11, by Calum G. Turvey and Hong Fu, who estimate the endogeneity between credit supply, credit demand in relation to agricultural productivity, and special expenditures on wedding and funerals.

The last two chapters before concluding provide comparative analyses between agricultural conditions in the 1930s and the modern era. Chapter 12 by Hao Hu and Feng Zhang examines cropland utilization and productivity, and Chap. 13, by Hao Hu and Funing Zhong, provides a comparison of the two eras in terms of changes in agricultural production including cropping structures, labor productivity, technology and so on. The book concludes with Chap. 14.

These chapters represent what we believe to be the initial round of analysis; no doubt more studies will be forthcoming. All told, the chapters either provide new insights into Republican-era agriculture or confirm existing observations. As is normal with an edited volume, the chapters represent the individual and independent works of the contributing authors. Our role as editors was to edit and streamline the papers into a reasonable flow, and we thank the authors for their cooperation throughout this process.

Nanjing, China — Hao Hu
Nanjing, China — Funing Zhong
Ithaca, NY — Calum G. Turvey

Acknowledgments

We have many people to thank. The empirical works provided in this book were the end result of years of preparation. Writing the actual chapters was probably the easiest part, so much is owed to the many scholars and students who contributed to the funding, preservation, and compilation of the data. Atop this list is Professor Sumio Kuribayashi of Tokyo International University (TIU), who as principal researcher led the initial team to preserve Buck's data. This was a five-year effort starting in 2002 and ending around 2007. Professor Kuribayashi arranged the first tranche of funding from the Grants-in-Aid Program of the Japan Society for the Promotion of Science (JSPS) under project # 15402020. Under his lead the original preservation study with Professors Hao Hu, Funing Zhong, Yingheng Zhou, and Qun Su from NJAU took root. This included an exchange program, with Takashi Osato from TIU and Wei Wei Liu from Nanjing Agricultural University being the first to partake.

Over the years that followed, further funding was provided by a grant from the National Social Sciences Foundation of China for the project Research on Construction of Agricultural Production System and Agricultural Products Consumption Pattern Characterized by Low Carbon Emission (Grant No. 10zd&031). It is also supported by a grant for the Construction of Buck's Survey Database, a Major Research Program of Humanities and Social Sciences, NJAU (Grant No. SKZD201201).

There are many students and faculty at NJAU to thank, but more than any we thank Weiwei Zheng and Minjie Yu who, as graduate students, undertook the tremendous burden of organizing dozens of undergraduate and graduate students to aid with the final digitization process. Both have contributed to this volume and now hold faculty positions in China.

From the Cornell University side, Calum G. Turvey would like to acknowledge the support provided by the W.I. Myers Endowment Fund that supports his exchange and a grant of $20,000 from Jeffrey Sean Lehman Fund for Scholarly Exchange with China to offset costs of travel and translation of manuscripts. He would also like to thank his colleagues in the Dyson School of Applied Economics and Management, the College of Agricultural and Life Sciences, and the Cornell College of Business for providing the flexibility and time to pursue works on Chinese agriculture and Chinese agricultural history. Turvey would also like to thank personally Professors Hao Hu and Funing Zhong for involving him in the project. Turvey contributed nothing to the ten years of data preparation, yet when he requested access to the data both Hao and Funing agreed, making him the only scholar outside of NJAU to have this privilege. This comes with additional personal thanks to Yu Yi, Weiwei Zheng, Minjie Yu, and Yongbing Yang, who, over the years, responded to many requests for data, data clarification, and overall support. Turvey would also like to thank Dizi Chang and Ziang "June" Cheng, who completed Cornell Master of Science theses on calamities and conflicts and credit demand respectively. The contributed chapters on these topics could not have taken shape if not for their efforts.

Finally, we would like to thank Elizabeth Graber, Commissioning Editor, Economics, at Palgrave Macmillan and her team, including Sujitha Shiney, Sophia Siegler, and Lucy Kidwell, for their technical advice, editorial contribution, and patience with our missing many deadlines.

Contents

List of Contributors

Hong Fu School of Insurance, Shandong University of Finance and Economics, Jinan, China

Hisatoshi Hoken Kwansei Gakuin University, Nishinomiya, Japan

Hao Hu College of Economics and Management, Nanjing Agricultural University, Nanjing, China

Qun Su College of Economics and Management, Nanjing Agricultural University, Nanjing, China

Mikio Suga Faculty of Economics, Hosei University, Tokyo, Japan

Calum G. Turvey Cornell University, Ithaca, NY, USA

Zhongwei Yang Business School, Nanjing Xiaozhuang University, Nanjing, China

Minjie Yu Business School, Huzhou University, Huzhou, China

Feng Zhang Jiangsu Academy of Agricultural Sciences, Nanjing, China

Weiwei Zheng Jiangsu Academy of Agricultural Sciences, Nanjing, China

Funing Zhong College of Economics and Management, Nanjing Agricultural University, Nanjing, China

List of Figures

List of Tables

CHAPTER 1

China's Agriculture in the 1930s: An Overview

Calum G. Turvey

1.1 Introduction

To understand China's agricultural economy today it is important to comprehend the country's history; yet few works have done so. The formation of the Republic of China in 1912 through the end of the War for Independence in 1949, the era of collectivization from 1949 through 1978, and the modern era from then on—these periods have all been characterized by distinct economic change. We cannot call them phases because by any measure the transformations were economic discontinuities. The warlord era following the fall of the Qing saw agriculture largely ignored until the 1921 famine that galvanized the China International Famine Relief Commission (CIFRC) to take action on its own account and initiate reforms in infrastructure, irrigation, wells, cooperation, and credit. It was not until 1928 or 1929 that the Nanjing government,

C. G. Turvey (✉)
Cornell University, Ithaca, NY, USA
e-mail: cgt6@cornell.edu

H. Hu et al. (eds.), *Chinese Agriculture in the 1930s*,
https://doi.org/10.1007/978-3-030-12688-9_1

headed by Chiang Kai-shek, started to promote agricultural reconstruction, but the inequities up to this point—particularly with respect to land tenancy—provided a foothold for Chinese communists under Mao Zedong and Zhu De. Despite the great floods and annexation of Manchuria in 1931, and the beginning of the Sino-Japanese War in 1937, the efforts of reconstruction continued, but ultimately could not withstand the forces of change brought about by the prolonged costs of the Second World War, the beginning of the War for Independence, the collapse of the national Chinese currency, and the defeat of the Kuomintang (KMT).

The second era of the twentieth century (1950–1978), this brought about a period of collectivization in which land ownership was abolished in favor of large, state-run cooperatives. This period saw the troubling Great Leap Forward between 1958 and 1962, during which resources were drawn away from agriculture, and then the Cultural Revolution between 1966 and 1976. The third era of the twentieth century (1978–2015) began with the household responsibility system in 1978 in which previously collectivized farms were broken down into small household units of perhaps one acre per family, and more progressive reforms, in terms of access to credit, market development, infrastructure development, mechanization, and so on, were put in place. Writing now in the fall of 2018, at a time which has seen new reforms to urbanization, land tenancy, and the formalization of land transfer centers and mortgaging, China is perhaps on the cusp of a fourth era starting with land reforms around 2015 leading to an era of agricultural commercialization and expansion.

The inspiration for this book is John Lossing Buck's 1937 volumes of *Land Utilization in China*, which are based on a large national survey of China's agricultural economy between 1929 and 1933. What Buck provides is an in-depth assessment of agricultural conditions and land use at a specific moment in time. Japanese data employed by Huang and Myers between 1935 and 1942, and the various surveys conducted by Buck in the development of *Chinese Farm Economy* in 1930, provide similar snapshots. The agricultural economist's job is to make sense of this data to determine the state of affairs with regard to general welfare and to decide what lessons might be learned to aid in the understanding of certain theories and, through that process, provide insights into effective policy. So, one side of the economist's task is to analyze data with regard to theory to see what emerges and describe it in the context of the time frame recorded. Another task is to place the economic conditions in a broader historical context. In the case of China, the evolutionary studies of Perkins and Elvin show how difficult that task is when data is so sparing. Another angle is how the economist should absorb the data and through what lens. Huang and Fei, for example, are anthropologists and examine economy via the local ecology, recording as they go how things such as social networks and kinship, community and culture might deviate from the neoclassical economic model. Huang introduces the notion of "involution" to describe certain economic behaviors of the peasant class that would appear absurd to some classical economists, while attaching classical economic profit optimization to the managerial class—two economically distinct cohorts at the same moment in time applying different sets of economic rules, each in the preservation of self-interest. The economist's task is to make sense of all this, not only in the moment, but in the broader frame of historical and future economic progression.

In the preparation of this volume we have been fortunate. After the 1937 publication of Buck's *Land Utilization in China*, it was thought that the actual household records were lost to the Second World War, but around the year 2000 they were discovered in the archives at Nanjing Agricultural University (NJAU). Starting in 2002 (and ongoing), the actual household records from the survey were copied, preserved, catalogued, digitized, and verified. This book draws upon these data. In this chapter I attempt to place Buck's work in a broader perspective. The microeconomic scope of each of the analytical chapters stand in context on their own. Here, I have elected to position the work in terms of equilibrium traps, of which involution may play a part, and see where this leads us.

1.2 Land Utilization in China

Core to this book is the discovery of Professor John Lossing Buck's discovery of the original survey spreadsheets used to compile *Land Utilization in China* (see Chap. 2). Which were stored, unnoticed in the archives of NJAU, scholars and students have spent nearly 18 years preserving, documenting, and finally digitizing the data for use in economic analysis. That is where this book comes in. The original three volumes of *Land Utilization in China* were compiled largely by abacus. The reconstructed microdata now permits investigation using sophisticated econometric and statistical techniques, not available at the time. The specific details of Buck's life in China need not be detailed here except to say that as an agricultural missionary with an agricultural degree from Cornell University, Buck arrived in China in 1915. Two years later he married the Nobel laureate Pearl S. Buck, and in 1921 they moved to Nanjing where he took up a professorship, with the task of establishing a program in agricultural economics. Over the years Buck conducted a number of farm surveys with the aim of determining household incomes, profits, productivity, land tenure, and so on. Earning an MS degree in agricultural economics along the way, these studies were compiled into his first book, *Chinese Farm Economy*, which was also used to satisfy the requirements for his PhD from Cornell in 1933. It was toward the end of this work that he was approached to make a much broader and systematic study of land use in China. This led to the publication of *Land Utilization in China* in 1937. An extensive study, it included 16,786 farms in 168 localities across 22 provinces, providing intricate detail on climate, land, crops, livestock, fertility, farm business, farm labor, prices and taxation, marketing, population, nutrition, and standards of living. Even today, Buck's book stands out as a standard reference of conditions on the ground in China.

1.3 Land and Labour in China

Our book, we believe, complements other great works on Chinese agriculture. The first of these, by R.H. Tawney, *Land and Labour in China*, in 1932, notes that "*the Chinese peasant is, by general agreement, a highly skilled farmer, who has achieved, in certain branches of his art, an extraordinary efficiency. But the centuries of tradition which have perfected his technique have also narrowed it*" (pp. 51–52). Tawney, drawing from multiple sources, including some of Buck's earlier works, noted that most farmers were semi-subsistent, with about 47% of crops grown consumed on the farm (generally the lower-quality and valued grains such as kaoliang), while 53% were disposed of beyond the farm, usually at low harvest prices. Although no data on credit was available to Tawney at the time of his writing, he noted that, observationally, indebtedness was extensive and often crushing. As the next harvest approached, savings and food stores were so depleted that the farmer was forced to borrow locally at usury rates to survive, and with the weight of credit upon him, a forced sale at harvest was required to free him from the burden. By 1930, efforts were underway to develop credit cooperatives and societies—with Buck playing a significant role—and to establish agricultural banks including the Farmers Bank of China (Fu and Turvey 2018), but these efforts were on a rudimentary scale in the 1929–1933 period.

Tawney viewed the land tenancy issue as less of an economic issue than credit. In the absence of a landed gentry, as observed in the feudal politic of Europe, the dominant form of contractual arrangement between a landlord and tenant was relatively formal, and if the Chinese farmer were to provide a list of grievances tenancy would not necessarily rise to the top of the list. Having said this, Tawney points out that in the North China Plain, the majority of farmers, perhaps as much as 75%, were owner-occupiers. In the southern provinces the tenancy relationship was quite different, with 90% of farmers being either tenants or some combination of owner-tenant. Not enough was known about tenancy in the northeast in 1931 and 1932, but Tawney was reporting that while tracts of newly opened lands were leased to farmers at reasonable rates initially, rental rates continued to rise thereafter. The point that Tawney was making was that despite numerous efforts to expand agricultural credit by the mid-Republican era, the limitations of formal credit institutions and cooperatives were resented by most farmers, while tenancy was resented by relatively fewer.

Neither production efficiency, credit, nor tenancy mattered, however, when the exogenous forces of war and famine ensured poverty entrapment. Throughout the Republican-era war, banditry, droughts, and floods were common hardships. And when they occurred few could carry the burden through to a next harvest, with famine and death a common result. Perhaps the most expansive chapter in this book (Chap. 3) describes the calamites, catastrophes, and conflicts faced by farmers during the 1929 to 1933 period. If under normal conditions between 87% and 90% of farm households could not afford to buy meat in a given year, one can only imagine the horrors that would follow a drought or flood or conflict.[1] Add to this the heavy taxation by local warlords or cadres or governors, in cash or grain, or the forced cultivation of non-edible crops such as opium to fill the military coffers, the best that the farmer in the Republican era could do was to survive at the margin of subsistence, consuming throughout the winter as little and as poor food as possible.

The labor economy was untenable. With population growth exceeding the pace at which new land could be brought into production, China, at

[1] Tawney (1932, p. 72 fn 1) citing F.C.H. Lee and T. Chin (1929) and Buck and Sever (1925).

the turn of the twentieth century, was a Malthusian nightmare. The land-to-labor ratio over the centuries not only left an abundance of labor idle, but nearly no opportunity for a livable wage market to develop. Tawney (1932, p. 135) points out that China's "*most serious defect—a very grave one—is that, owing to its abundance, human labour is cheap, with the result that the introduction of machinery, which, had labour been dearer, would have taken place long ago, has been discouraged.*" Even in the best of times labor constraints for the profit-minded firm would never be binding, and in not being so the wage rate would never reflect its true marginal economic value. Instead, the wage rate would be set by the most desperate of souls, requiring a pittance for the most meager of food. Unconstrained by supply, labor demand would be almost perfectly elastic at the lowest of survivable wages—or less. And this was in good years. In bad years, with risks shared across small and large farms, the demand would fall as supply would rise, resulting in even more desperate conditions.

1.4 The Chinese Peasant Economy

Another classical study is Ramon H. Myers' *The Chinese Peasant Economy*. Myers based his analysis on data collected by the Japanese and the South Manchurian Railway of villages in Hebei and Shandong between 1939 and 1943. Myers set forth two hypotheses to explain the agricultural conditions of China at the time. The first was the "distribution theory," which held that such a large portion of income was taken from farmers in rents, high interest rates, taxes, and unfair terms of price exchange that they were left with little surplus to improve or enlarge their farms and raise their living standards. The consolidation of wealth amongst the wealthy allowed them to purchase more and more land, leading to increased inequality; this gentry class also took a larger and larger share of surplus, leaving little beyond subsistence for farmers to make improvements beyond traditional levels. The second theory was referred to as the "eclectic theory." This theory held that farmers were poor, not only because of taxes, and high rent and interest rates, but also because production was depressed in the first place. The causes of this were inadequate organizations of farms, poor infrastructure, inadequate government support, and insufficient amounts of basic inputs with which to improve yields. Which of these pathways were true will likely never be resolved because it seems in reality that the truth is either somewhere in the middle or some combination of the two. Nonetheless, Myers concludes that agricultural problems facing

Chinese farmers in the Republican era were driven by persistent population expansion that led to overcrowding and an increase in landless farmers; that there was great neglect of economic infrastructure including transportation, irrigation, and flood control systems. These were brought about by incompetence and grift. Finally, there was the landlord class that which "rack-rented" the peasantry and tried to accumulate more land for itself (Myers 1970, p. 23).

1.5 Agricultural Involution

Philip Huang's *The Peasant Economy and Social Change in North China* is an expansive study based on data collected by Japanese occupation forces and researchers in Hebei and Shandong between 1935 and 1942. Huang notes that over time researchers from different branches of agricultural and development economics have taken competing views of what comprises a peasant economy. The first view comes from the "formalists." This branch looks to classical economic for solutions and sees the peasant as an economic man who is motivated by profit maximization and optimizes toward an equilibrium subject to the constraints faced. The second view, termed "substantivist," is driven largely by economic anthropologists who stress social relationships, including the strong bonds of kinship, reciprocity, and other-regarding preferences, which, when combined with risk avoidance and safety-first choices, result in a subsistence ethic that does not fit neatly into the classical arguments of utility maximization. The third view, drawn from conventional Marxist arguments, is held by the "feudalists," who cast the peasant as an exploited cultivar whose surplus goes to supporting a ruling gentry and state structure. To the feudalists, the peasant economy has land at its core, with the beneficial economic argument driven by contracts between landlord and tenant and bargaining power.

Huang refuses to peg the Chinese peasant to any one of these categories and makes the case that the agricultural economy of the day was a coexisting amalgam of the three. Like Myers (1970), Huang used data gathered by Japanese researchers linked to the South Manchurian Railway Company between 1935 and 1941 in Shandong and Hebei. This is probably the correct approach. Even Myers (1970) initially proposed, and then rejected (p. 292), the argument that it was economic growth rooted in socioeconomic relations which determined the distribution and production of rural wealth.

Huang introduced the idea of "involution" to explain the economic conditions under which a managerial farm of means with skilled labor and access to capital produced greater output per labor-equivalent to smaller farms, but an equivalent total output per mou (1 mou = 1/6th of an acre).[2] Huang explains that small farms "involute" by applying more labor than was optimally necessary and at the cost of lower and diminishing marginal productivity. The reason for doing this was to keep excess family labor busy. But in doing so the total product per unit area would be equivalent to the more industrious farm that employed more labor animals and fertilizers and lower amounts of "wage" labor. It would not be unusual for a farm to select more labor-intensive crops, rather than more profitable crops, in order to minimize slack in household labor utilization.

Were the managerial farms able to hire more labor or local infrastructure improve enough to promote industry, the familial supply of labor could be reduced and the economic costs of involution could be reduced or averted. However, Hsiao-Tung Fei, writing of a village in Jiangsu province circa 1935 noted that "*changes in occupation is difficult and even change of crop seldom comes to the mind of villagers. Thus the structure of production is a rigid one and does not react elastically to the demand of the market. When changes take place they are gradual and far reaching* ... (on structural changes in the silk industry) ... *To bring about a change in the industry, special knowledge and social organization are needed. All these factors delay an immediate and automatic adjustment of supply and demand*

[2] The term "agricultural involution" was coined by Clifford Geertz (1963), who in turn borrowed the term "involution" from the anthropologist Alexander Goldenweiser (1936). Geertz observed what he thought a paradox; that in Indonesia with distinctly separate wet-rice land and sugar-cane land, there was a measureable increase in rice yields as the rice paddies (*sawah*) got closer to the sugar fields, and an increase in sugar cane as the plantations got closer to the rice fields. The rise in rice yields at this point of convergence tended to increase in a way that offset the loss in *sawah* displaced by increased plantings of sugar. The only other exceptional variation was that in the region of convergence there was greater population density, and for reasons inexplicable to Geertz, both crops flourished, as did the local population. Economics alone could not explain the higher yields. Geertz saw a cultural pattern that was neither stable nor transformative. In other words, despite a low land-to-labor ratio the excess labor did not shift toward more industrial uses (for example). Instead, the cultural dynamic was organized in such a way as to check development, or at least hinder it. From the anthropological viewpoint, once the cultural pattern was established, further change was inhibited by the tenacity of the social bonds. Instead, as population rises relative to land, and the marginal productivity of each labor unit diminishes, the intensity of labor would find opportunities internal to the system that would permit the total product of the region to either increase with surplus, or at least remain constant.

in the rural economy ... (on labor mobility) ... *Change of occupation in the village is more difficult than reform of an existing industry. No serious attempt has been made to find out the possibility of introducing new industries to the village ... those that go out to find new occupation are mostly young girls who have not yet entered into a fixed social place in the community* ... (on kinship) ... *The reaction against disruptive forces in social stability become a force to counteract the present mobility*" (pp. 260–262).

This passage provides an interesting example of involution in the Republican era. It illustrates the immobility of labor, the inertia in capital allocation, and the general inelasticity of change. One argument put forth is the so-called "Needham puzzle," which centers on two fundamental questions regarding China's economic and industrial evolution[3]: First, why, historically, had China been so far in advance of other civilizations; and second, why isn't China now ahead of the rest of the world? By the fourteenth century China was cosmopolitan, technologically advanced, and economically powerful, so much so that in relation the West was essentially agrarian, poor, and underdeveloped. Over the following centuries this situation reversed. Dwight H. Perkins, for example, notes that starting around the fourteenth century, there were no dramatic changes in farming techniques or in rural institutions, even though there were infrastructure improvements to dykes, canals, and irrigation systems, while Mark Elvin notes scientific achievements in propagation and breeding. But it is likely more complex than that. Common arguments center on a failure in the demand for technologies as the population, and labor pool, grew faster than agricultural output. With abundant labor there was no need to introduce labor-saving technologies. On the other hand, there was a failure to supply technologies. For example, Justin Yifu Lin argues that the Needham puzzle arose from institutions in China that created bureaucracy and customs which no longer rewarded innovation. This was particularly acute in the post-Confucian era following the Sui (589–617) and Song dynasties (960–1275) when the enlightenments of science were replaced by rote memorization of Confucian scripts to pass the civil service examinations. To achieve high status in Chinese society it was knowledge of the scripts, and not the perpetuation of what underlay them, that was most important. The best and brightest were therefore forced to forgo investigations into science, technology, and engineering.

[3] Reference is to British sinologist Joseph Needham, who wrote *Science and Civilization in China*. Volume 1. Cambridge: Cambridge University Press, 1954 and multiple volumes thereafter. See Simon Winchester's biography on Needham (2008).

In many respects, Lin's argument supports earlier views by Tang (1979) that feudalism in Europe likely led to a dampening of agricultural innovation, while at the same time in China, feudalism had long disappeared in favor of private ownership. In Europe, as feudalism disappeared in the seventeenth and eighteenth centuries, just at the time the Industrial Revolution took off, there was every incentive to accumulate and apply scientific agriculture methods to the cultivation of food for an ever-increasing population. Meanwhile, in China, post-Confucian conservatism and the laser focus on civil service examinations stymied scientific innovation so that the maximum potential product diminished in proportion to the population.

However, involution was not the only source of labor inefficiency. Huang also argued that in areas where a labor demand and wage market existed, the allure of wages would be highest at the same moments that farm labor demand was at its highest, for example, harvest time. Smaller farms would deploy family labor in the wage market, reducing the amount of labor available when it was needed most and had its highest marginal productivity. And so the presence of a wage market could reverse involution, but, in a paradoxical way, bring about a different set of labor inefficiencies.

Huang's assessment of agricultural conditions in the Republican era rested largely on the land-to-labor ratio; he noted that the ratio increased as farm size grew from small peasant farms to larger managerial ones. At critical times throughout the cropping year when idleness is unprofitable, he argued, the household land-to-labor ratio would become involuted, with excess familial labor from small farms moving toward larger farms with deficits in familial labor. The contemporaneous measure of land-to-labor would temporarily rebalance, with the land-to-labor ratio for small farms increasing and for large farms decreasing. With small farms outnumbering larger farms by a significant number—61% being medium or smaller, 19% being large or larger, and 20% being medium-large[4]—larger farms having more persons per household than smaller farms,[5] and land per family member being lower for small farms than large farms,[6] the

[4]Buck (1937a), Table 6, p. 271.

[5]Across China, the average number of persons per household was 6.2. Small households had 4.4 members, with large and very large farms having 8.3 and 10.1 persons respectively (Buck 1937a, Table 14, p. 278). This pattern also appears to hold in Huang's Tables 14.1 and 14.2, pp. 250–253. See also Myers (1970), Tables 17 and 18, pp. 132 and 133.

[6]Across China, land per person (acres) was 0.21 for small farms and 0.77 for very large farms, with an average of 0.43 (Buck 1937a, Table 15, p. 279).

absorption rate of labor by large farms would not always be complete, leading to an excess supply of labor for small farms. This excess labor would substitute for animal labor.[7]

In *Chinese Farm Economy*, Buck (1930) provides a comparison from his farm surveys conducted across China in the 1920s.[8] In China, he notes, the chief factor in agricultural production is labor, whereas in the USA capital investment in improved tools and farm machinery plays an important part. On labor versus technology one hour of labor produced 1.1 kg of corn in China and 45.5 kg in the USA, for wheat it was 1.6 kg/man hour in China and 39.4 kg/man hour in the USA, and for rice 2.2 kg/man hour and 18.7 kg/man hour in China and the USA respectively. The multiples of labor required in China per man-equivalent in the USA to produce one hectare of crop was 5.61 for cotton, 5.83 for potatoes, 14.11 for corn, 23.1 for winter wheat, and 7.09 for soybeans.[9] Because of the land/capita problem the technologies adopted in the USA would be impracticable in China. The fixed costs of a tractor and gang plow to cover 60 hectares of land cost about $4.75/ha with operating costs of $10.43/acre in the USA. In China the costs of plowing a field with water oxen was about $4.00/ha.[10] Buck notes that "*because of the dense population, the Chinese farmer is doomed and all that can be done is to make the most out of an unfortunate situation*,"[11] and later "*The remedies for this too small size of farm business are difficult to find … As China becomes modernized, it is inevitable that industries will develop and a certain number of the country people be absorbed into them. Yet it can scarcely be hoped that sufficient numbers of them be absorbed as to relive the present agricultural situation very much. The best future solution of the problem seems to be in some method of population control, and the best immediate solution, more intensive methods of raising crops and the growing of crops that produce more food per unit of land. Such productivity, however, will also be useless if* [the] *population continues to grow*."[12]

[7] On average across China there were 4.9 labor animals per farm. Small farms averaged 2.6 animals while very large farms averaged 6.7 labor animals (Buck 1937a, Table 13, p. 277).

[8] Buck, J.L. (1930) Chinese Farm Economy: A Study of 2866 Farms in Seventeen Localities and Seven Provinces in China. University of Chicago Press, Chicago Illinois, pp. 147–148.

[9] Buck (1930). pp. 230–233.

[10] Buck (1930). p. 315.

[11] Buck (1930). p. 314.

[12] Buck (1930). p. 424.

It appears that Huang's conclusion on the existence of agricultural involution in China has some merit. As Geertz explains it, agricultural involution is a phenomenon in which multiple ecological, anthropological, and economic forces conspire toward increasing complexity in order to sustain or maintain a given state. Buck (1937a), for example, reports that that there is no difference in the yields of farms across groupings from small to large, noting that it "*is sometimes assumed that yields on small farms are larger than those on large farms because of the supposed greater intensity of culture on the small farms*" (p. 273). But, as noted, this is not observed. The anthropological/ecological assessment of involution would suggest that society adapts to the realities of intense population pressures by placing more economic value on small increases in the average product of labor even when the marginal product of labor is declining or even negative.

Can involution explain the state of China's agricultural economy in the 1930s? An involuted economy is one that is self-organizing. In the absence of complete capital markets, efficient agricultural/non-agricultural wage signaling, and public/private safety nets, the exogenous world of conflicts and catastrophes can create an impenetrable barrier to change. Despite the opportunities to acquire productive assets or otherwise engage in innovative and entrepreneurial activities, the frequency of, and uncertainty about, some future calamitous event causes the community to become insular and inward looking, retrenching to a safety-first world and a permanent state of structural and technological inertia. This results in "*peculiar pattern of changeless change …*" with a community not comprised of "*haves*" and "*have-nots*," but one of shared poverty comprised of "*'just-enoughs' and 'not-quite enoughs'*" (Geertz 1963, pp. 96–97). Consequently, the bonds of kinship and community strengthen as do mutual aid, informal familial lending, and a deep-rooted commitment to reciprocity. This moves the fabric of social preferences away from self-interest and in favor of inequity aversion and other-regarding motives; members within a social network care not only about their own material payoffs, but also about the distribution of payoffs of others. For the most part Huang attributes involution to the lower income peasant farmers, adding that larger farms will generally optimize under a profit-maximizing motive. With shared social preferences and kinship it is not surprising that clusters of farmers exhibiting some form of involution would appear across rural China.

1.6 Equilibrium Traps

If the concept of agricultural involution holds true, then this might provide some insights into a class of economic models referred to as equilibrium traps. These are precursors to the more recent notion of poverty traps. We will, with some trepidation, use the terms equilibrium trap and poverty trap with some interchangeability since both are rooted in the neo-Malthusian approach to equilibrium by Nurkse (1952) and, earlier, Malthus (1888). Three typologies identified in Carter and Barrett (2006) are the chronically poor, transitory poor, and never poor. The first two can be further identified by structural and stochastic measures. Are the chronically poor bound to persistent poverty because they could never obtain land and/or productive assets in a way that provides them an escape; or are they subject to the consecutive random shocks that have plagued China's agricultural economy for thousands of years? Are the transitory poor simply more resilient with savings and consumable assets available to preserve land and productive assets so that poverty is but a temporary state? Finally, in the longer run are there are dynamic adjustments that can determine whether the asset poor can move above an asset threshold with improved returns that will, in probability, raise the household above the poverty threshold?

In terms of the multiple equilibria differentiated by asset or income class, the chronically poor are most likely to face involution because they make up the most labor-intensive cluster. The transitory group, which can move in and out of poverty, may share similar social preferences as the chronically poor, but are less likely to face involution because their transitional poverty states are determined more by stochastic forces rather than structural deficits, except in extreme situations. Where this might differ from the asset poverty trap model, is that involution reflects a semi-permanent state that establishes at least one stable equilibrium in an agricultural economy with multiple equilibria. The asset-rich wealthier farms will also constitute a stable equilibrium because, being resilient, they are virtually immoveable even under stochastic conditions, and will recover fairly rapidly. The transitory group constitutes a third equilibrium but this is an unstable equilibrium in the sense that this group will be forced toward the asset-poor poverty equilibrium when times are bad, while voluntarily gravitating toward the asset-rich equilibrium when times are good.[13]

[13] Of course, this approach to multiple equilibria could be broken with the political risks facing the asset rich during the Republican era, when communist forces created soviets in which deeds of land were seized from the asset rich and redistributed to the asset poor in an

1.6.1 Low-Level Equilibrium Traps

Several demand–failure models have been developed to explain the lapse in agricultural innovation in China.[14] Prominent among these is the model put forth by Elvin, but before this were those by Nurkse and Nelson. Nurkse (1952) makes the argument that balanced growth rests ultimately on the need for a balanced diet.[15] The imbalance results, at least in part, to the inelasticity of demand for consumables at low real income levels, so that almost all goods are seen as necessities. Thus begins the circular relationship in low-income economies that the inelasticity of demand leaves little capacity to save, and thus the capital to invest, and thus to low productivity. The lack of buying power impedes any incentives to invest in a diversified industrial base that would ordinarily provide complementary goods and services so that the new entrepreneurs become each other's customers and slowly extract themselves from the deadlock of a low-level equilibrium trap.

Drawing on this, Nelson (1956) built a macro dynamic growth model with multiple equilibria, one of which can result in a stable low-level equilibrium trap.[16] Equilibria are established when the population growth rate $\frac{dP}{P}$, under an accumulation of capital, K, equals the growth rate in output, $\frac{dY}{Y}$. Capital is comprised of produced goods including productive assets, K', and land available for cultivation, L, which Nelson allows to be perfect substitutes. Even if capital inputs stagnate, capital can still be increased by cultivating more land, but as more land is brought into production there is an increasing difficulty with regard to whether new lands would be of equal productivity to previous lands. Capital formation, changes in population, changes in output, and the social, political, and economic organization of the economy ultimately determine various equilibria where changes in population equal changes

attempt to achieve a single stable equilibrium with mutual aid, and a collective labor force. Ultimately, asset reallocation would dissolve the social fabric that bound involution to a distinctive asset-poor poverty cluster, so it is unlikely that involution would be observed in the mid-Republican soviets.

[14] This section draws on Fu and Turvey (2018), Chapter 3 "Low-Level Equilibrium and Fractional Poverty Traps."

[15] See Nurkse, R. (1952).

[16] See Nelson, R. R. (1956).

in output. Changes in population are bounded by the maximum biological rate and it is assumed that this arises only after a period of time at which per capital output was substantially higher than a subsistence rate. The boundary to the low-level equilibrium poverty trap is distinguished by the boundary of subsistence; that is, (typically underdeveloped) economies in which output per capita is at or below subsistence, versus those (typically developing or developed) in which output per capita is above subsistence. The low-level equilibrium trap relates to low-income/low-technology states and is a stable equilibrium—an equilibrium that persists—when the population growth rate intersects the output growth curve from above.

The situational forces in Nelson's model are depicted in the three panels of Fig. 1.1. The x-axis is represented by per capita output, $\frac{Y}{P}$, which is also translatable to per capita income. At $S = \frac{Y}{P}$, is per capita income at subsistence. The determinative economic forces are the $\frac{dP}{P}, \frac{dY}{Y}$ curves. Population growth is increasing during the capital formation phase, but diminishes as per capita output increases, to an almost zero growth rate starting around point B. The output curve is determined by endogenous and exogenous social, political, resource-based factors, and so on, which are uniquely determined on a region-by-region or country-by-country basis. Equilibrium is established where $\frac{dP}{P} = \frac{dY}{Y}$, at which, at least over the short run, changes in aggregate per capita output and income just equal population growth, putting the economy in an exact balance.

The top panel of Fig. 1.1 shows three equilibria at points A, B, and C. Initially the population growth rate exceeds the growth in output as it crosses the subsistence level at point A. This continues over the range A to B, crossing the output curve at point B. Between A and B per capita output is above subsistence, but still deficient and requiring greater deployment of productive capital and/or land to close the deficit. The deficit is closed at point B, which is an unstable equilibrium. Between B and C output increases exceed population increases, which would result in economic surpluses. The economic value of these surpluses eventually recede to a new and stable equilibrium at point C, with per capita income at $\frac{Y}{P}^{**}$. Nelson coins the equilibrium at point A as a low-level equilibrium.

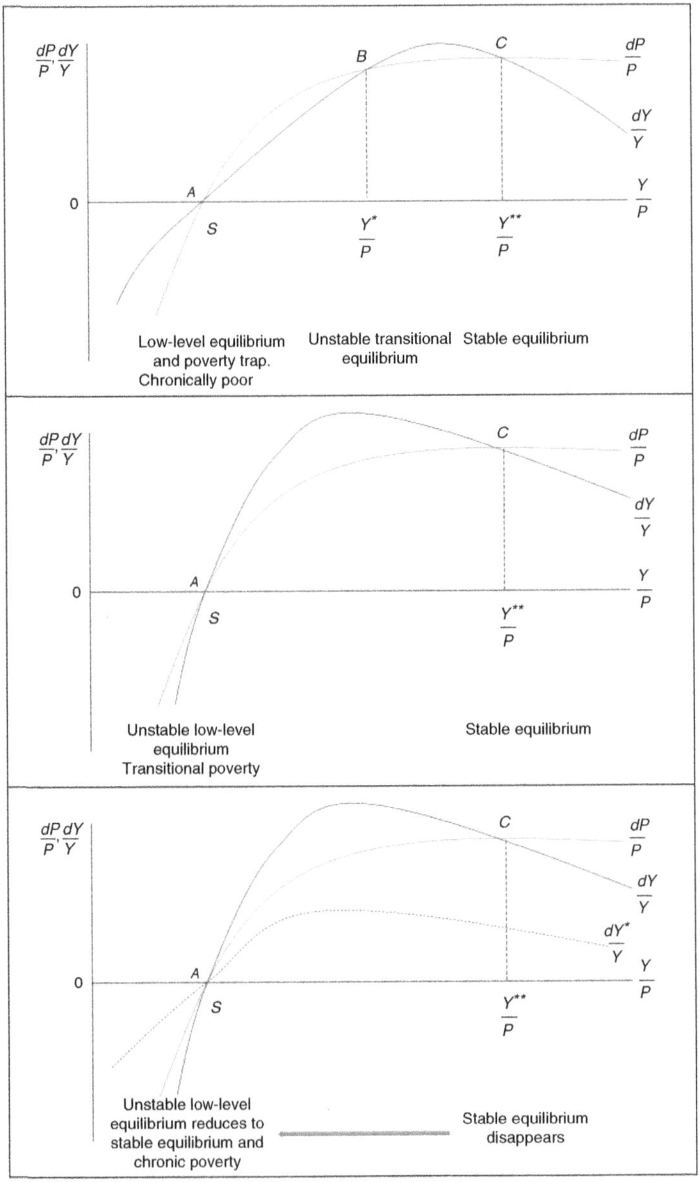

Fig. 1.1 The low-level equilibrium trap

At this point changes in aggregate output equal changes in population at the level of subsistence. In order to break the equilibrium there must be a structural shift in the deployment of capital, including the addition of cultivated lands or the adoption/invention/importation of productive capital. This low-level equilibrium creates a poverty trap. The strength of this poverty trap depends upon the ability of the population to mobilize capital so that capital per worker increases. If the pace is rapid then the equilibrium at point B will shift to the left, and if slow to the right.

The middle panel in Fig. 1.1 reverses the initial conditions. Here, $\frac{dY}{Y}$ crosses $\frac{dP}{P}$ at S, from below. Point A is an unstable equilibrium and is transitory. Capital formation will adjust relative to population growth with near continuous and accumulating economic surpluses until a new and stable equilibrium at point C, bypassing the unstable equilibrium at point B in the top panel entirely. However, for agrarian economies in which the cultivation of new lands substitutes for productive capital, Malthusian forces will come into play. Initial cultivation would be in rich and fertile valleys but as population increases later cultivation would include lands of lower quality with diminishing marginal productivity, requiring more intensive use of labor in order to obtain the same level of output. In the absence of technological innovation and/or the deployment of productive capital, the productivity of new lands would decline at an increasing rate, so, with continued population growth, the aggregate land-to-labor ratio would continue to fall. Consequently, the $\frac{dY}{Y}$ curve twists clockwise so that only one stable equilibrium remains at point A. In the absence of any external force to push complementary technologies to offset the diminishing marginal product of land, the economy would fall into a poverty trap, and that poverty trap would persist.

In terms of the development of China's agriculture economy throughout the Republican era this model provides some important insights. Point B, as mentioned, is an unstable equilibrium. If capital per capita decreases then the economy moves from point B toward point A, but if it is sustained it moves toward point C. Somewhere around 1400 AD China's economy was probably to the right of point B. But for some inexplicable reason, often referred to as the Needham puzzle (see Lin 1995, 2008), innovation and capital formation slowed, or stopped altogether, at least to the point where changes in output could not keep pace with changes in population. Much is made of the decline in inventiveness in China relative to progressive inventiveness in Europe, resulting in Europe shifting from an unstable equilibrium at point B to a prosperous stable equilibrium at point C, while China shifted from the unstable equilibrium at point B toward the unstable equilibrium at point A. This has led some scholars to question whether, at the turn of the twentieth century and into the Republican era, China's agricultural economy was in a low-level equilibrium trap.

Once a low-level equilibrium is established, escaping the trap is not immediate. If the trap begins with a change in population intersecting the change in output from below then the trap will remain until the change in output again exceeds the change in population. This can come about by an injection of capital into research and development or infrastructure, or the development of new markets to spur on demand. But until that happens, the output per capita is declining and in the Malthusian sense this can also come about by declines in population due to emigration and famine.

The crucial insight from Nelson's Malthusian trap model is representation of the existence of persistent poverty traps that arise from the relationship between changes in population and changes in output or income. In critique, Hagen (1959) observed that in no Old World country did the population growth rate ever exceed the rate of growth in aggregate output. Perhaps, but in the identification of multiple equilibria it is the characteristics of the respective clusters that matter, and not so much aggregate output. In fact, Hagen does note that in the case of China's population growth the introduction of sweet potatoes, peanuts, and early ripening rice allowed the cultivation of lands that could not previously support the population (p. 312). As part of the involution paradigm, agricultural output could be sustained with declining marginal labor productivity and a diminishing land-to-labor ratio if that labor is put to use in improved cultivation practices, or if new labor-intensive crops are introduced. Nonetheless, China seems to stand out in being a country that over time expanded lands while not pushing the technological frontier to keep pace. As mentioned, it is unlikely that that there was no technological advances in China during this agricultural phase, albeit not at the pace observed in Europe or North America.

1.7 High-Level Equilibrium Traps

Nelson's model is essentially a static short-run model. As the technological, economic, and political environment changes then so too would conditions leading into new equilibrium traps, or exiting existing equilibrium traps. An alternative model is Elvin's proposition of a high-level equilibrium trap (1973, 1984, 1996; with Sinha (1973)). Elvin (1972, 1973) uses the term "high level," rather than "low level," to describe an equilibrium in which all innovations toward a maximal level of agricultural productivity have been exhausted at both the intensive and extensive margins. A low-level equilibrium assumes primitive technology so that output can be increased by simply applying more labor into the market. Elvin notes that productivity per mou in 1368 was about 140 catties of grain (1 cattie is about 500 grams), rising to 224 by the 1770s. It then fell to a little above 200 catties, rising to 240 catties by the mid-1800s. Connecting a line between the 1368 and 1850s high approximates the potential output available given the levels of technology available. That for almost 500 years actual production did not (with records available) breach this potential led Elvin to reconsider the underlying dynamics and diffusion of output per capita, while still recognizing that China had agricultural practices that were more advanced in terms of pre-modern technology than other countries.[17]

Figure 1.2 reproduces the Elvin–Sinha conceptual construct of the high-level equilibrium trap. The curve 0AT represents the maximum potential output that could have been obtained had the economy employed the maximum capital per unit of labor available. A point of departure from Nelson is that Elvin believed that the Chinese had the human capital available to innovate or expand the use of existing technologies, but for some inexplicable reason chose not to use them.

The line 0S represents subsistence. The x-axis is indexed to labor, which is assumed to be proportionate to population, and consumption demand that also rises in proportion to population. The gap between potential output and subsistence is measured by the vertical difference of which the lines AC and FH are examples. The area below subsistence, for example, CD and HI, ultimately experiences unsustainable famine conditions. Assuming that time coordinates with population along the x-axis, Elvin argues that instead of exploiting all capital available, farmers selected

[17] See Sinha, (1973). fn 62, p. 19.

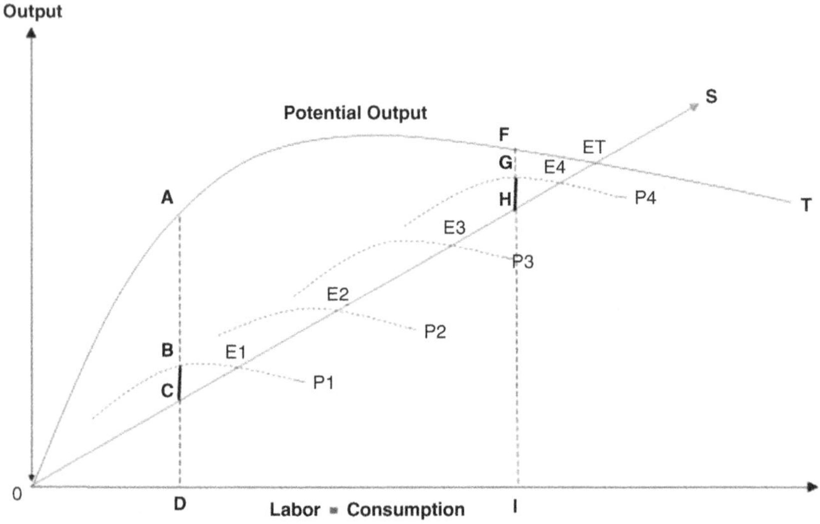

Fig. 1.2 The high-level equilibrium trap

a smaller subset and in a piecemeal fashion. This gives rise to a number of low-level equilibriums at E1, E2, E3, and E4. These equilibria are not traps, at least in the sense of Nelson, because as population rises relative to the production possibilities of the engaged technologies, there are technologies and capital in reserve that can be positioned. Thus, as the surplus BC along the production possibility P1 diminishes toward the low-level equilibrium at E1, that equilibrium is avoided by moving to a new production possibility at P2, and then P3 and P4.

Ultimately, and across generations, the inventory of capital and innovation made available by 0AFT diminishes as population pressures, combined with the diminishing marginal productivity of new lands and a decreasing land-to-labor ratio, take their toll. Elvin defines the equilibrium at ET, where the potential output curve crosses the subsistence line, as a high-level equilibrium trap. In the absence of a revolutionary investment in capital per labor and/or capital per land, the economy can no longer advance and—as with Nelson—the change in output falls below the change in population. If the trap persists the economy moves toward increasing destitution with all its concomitant effects.

In the China context, Elvin notes that the late traditional economy, starting perhaps in the fourteenth or fifteenth century (dates that correspond with the Needham puzzle), was incapable of changes through internally generated forces, which focused on China's obsession with male heirs and family lineage that led to outcomes of population increase whether aggregate output could sustain it or not. As the population rose so too did the pressures on arable land, which was in fixed supply. In early phases of population growth the population could be supported by expanding agricultural production, but the land expanded into likely had increasingly lower productivity. Eventually, the combination of decreasing labor productivity (involution) on a diminishing land base resulted in a population that could exist only at the margins. As populations increased, the cost of labor fell relative to investments in capital, removing any economic incentives to develop labor-saving technology. In addition, as poorer quality land came to be cultivated the rental value of that land—its marginal value—also fell so that the demand for technology also dropped. As lower productivity land was being brought into production for an increasing rural population, the retentions held back for household consumption as a proportion of total output would also increase. Thus, the proportion of output that made its way into urban and industrial centers would decrease relative to population increases there. Prices at the market centers would then have risen, causing a decrease in real wages, which would in turn reduce demand for other non-food items. And so it went in a spiral that removed economic incentives to expanding entrepreneurship and technological, scientific, and economic innovation. With a stagnation in innovation an equilibrium trap would have been inevitable as the rate of increase in potential output would fall below the rate of increase in population and labor. Instead, farmers would have adopted one scale of technology, the efficacy of which diminishes as population increases. Then a higher, level of technology would be adopted and used until it can no longer support the population. Even good public works would have diminished as taxing lower output per capita became increasingly more regressive. In this model the decreasing land-to-labor ratio coupled with a lack of demand for new technology would lead to a high-level equilibrium trap.

Elvin's model is speculative and not all scholars are in general agreement. Tang is skeptical of it, citing Buck's (1937a) survey to suggest that output was below maximum potential, so at least as late as the mid-Republican period China might have escaped the trap. Lin (1995, 2008) is not in agreement with Elvin's case for a high-level equilibrium trap. Lin's argument is that regardless of location there will be distributions of innovation recognized by chance and experience, or developed by enlightenment and experimentations suggested along the way, and that technological innovation requires a diffusion process across time. This would be true in China as well as Europe. Huang agrees somewhat with Elvin's notion that a failure to continually adapt new practices or develop new technologies can lead to periodic traps in which a failure to innovate combined with population pressure can lead to a high-level equilibrium trap, but he dismisses the idea that population alone is the driver, and that agricultural conditions cannot be understood in the absence of interdependent relationships between the natural environment and the sociopolitical order.[18] The point of departure may lie in the concept of involution. To Huang (it appears that) involution is a consequence of exogenous forces, including population growth and land quality, which drives diminishing returns to labor, whereas to Elvin (it appears that) involution is a consequence of endogenous forces that lead to a choice not to innovate, until it is necessary to do so when population growth absorbs any surplus.

1.8 The Malthusian Trap

I have previously referred to the various forms of equilibrium traps as being neo-Malthusian. Malthus, writing at the turn of the nineteenth century, was heavily influenced by the writings of the French missionary P. DuHalde in 1742 and 1772 on China's economic and social conditions, including its population. In DuHalde's 1772 report he describes the land pressures then already existing on the plains that "*neither hedge nor ditch is to be seen, and but few trees, so much are they afraid of losing an inch of ground.*"[19] The Malthusian trap derives from the mathematical reality that little wealth can be accumulated when the rate at which the population increases is greater than the rate at which new lands can be cultivated. By the late 1700s, and for hundreds of years before that, the terracing of even the rockiest of mountains suggest that land–population pressures persisted across China. As the population increased, the land-to-population or land-to-labor ratios decreased rapidly, thereby tethering the country's population to the edges of subsistence: "*It cannot be said in China, as in Europe,*

[18] See Huang (1985). pp. 182–184.
[19] DuHalde, P. (1772), p. 211.

that the poor are idle, and might gain a subsistence if they could work. The labours and efforts of these poor people are beyond conception. A Chinese will pass whole days digging the earth, sometimes up to his knees in water, and in the evening is happy to eat a little spoonful of rice, and to drink the insipid water in which it was boiled."[20] In times of famine the toll on human life was so great that the pressures on the land-to-labor ratio found some reprieve, but as good followed bad, the population pressures again took hold.

The conditions described by DuHalde and Malthus appear to confine or trap the agricultural economy in the neighborhood of point S in Fig. 1.1. As population growth exceeded growth in production due to population rises and the diminishing marginal productivity of new lands brought into production, conditions would fall below and to the left of point S, until the onset of some calamity temporary shifts conditions above and to the right, only to be reversed again. For S to be stable over time, it is necessary that the diminishing productivity of newly cultivated land be offset by gains in human capital to drive as much productivity from the land as possible. In this sense the Malthusian poverty trap is not a static condition but an intertemporal dynamic condition that stochastically oscillates around a low level of subsistence. Fu and Turvey (2018, Chapter 3) have referred to this as a "fractional poverty trap" with per capita consumption being ergodic, or mean reverting. If industrial economic growth flows from the wealth derived from a flourishing agricultural economy, it is not surprising that the Needham puzzle is rooted in China's agricultural development. Persistent poverty in agriculture ultimately deprives the general economy from much needed savings and capital, while an abundance of labor suppresses the need to innovate.

1.9 Nutrition

Perhaps the greater challenge facing the Chinese farmer in the 1930s was the adequacy of food and the balancing of nutrition in calories and protein. Nutritional security is determined by a number of interconnected factors, some endogenous to the household, others exogenous. Of the exogenous factors the greater impacts are from drought, floods, war, civil war, banditry, and diseases such as typhus. From exogenous factors the death toll in China has been staggering. Tawney (1932, p. 76), for

[20] Malthus, T.R. (1888). p. 105.

example, reports that the great famines of 1849, 1878–1879, and 1920–1921 saw 13.75 million, 9 to 13 million, and 500,000 die respectively. Famine in Shaanxi from the 1928–1930 drought took 3 million lives from starvation and in Gansu between 1926 and 1930 nearly a third of the population perished from famine, civil war, banditry, and typhus. There is evidence that localized rebellions and banditry are symptoms of, and not a cause of, food insecurity, although the fatal outcome is much the same. In earlier periods in China's history—between 202 BC and 8 AD, and later around 604 AD—China saw the development of granaries to ensure that surpluses from good years could be distributed in lean years without the price gouging that merchants and usurers would apply (Fu and Turvey 2018, p. 126). These granaries were eventually used to support the Green Sprouts Policy around 1069–1076 AD, which offered joint-liability mutual loans to farmers that could be used to smooth consumption between planting and harvest (Fu and Turvey, pp. 132–133). Huang (p. 109) describes remnants of these granary models near Jinan in Shandong during the late 1930s as a "society of wheat buns" that charged members 20 cents per month for the right to obtain a secured source of wheat at fair prices. Nonetheless, the constant blight of famine finally pushed structural reforms and reconstruction across China: from crop improvements and digging wells, to agricultural credit and infrastructure (Fu and Turvey 2018, Chapter 7). Chapter 3 in this volume details the calamities and conflicts facing rural China during the 1929–1933 period of Buck's land utilization study.

But what about in normal years? The typical farm household in 1930s China grew food on the farm, stored according to customary weights and measures enough grains to last through the next harvest, and sold the remainder as a cash crop on the market. In Fei's (1939) Jiangsu village study, for example, the rice staple was stored in amounts of 250 kg for an "old man above 50," 210 kg for an "old woman above 40," 330 kg for an adult male, 250 kg for an adult woman, and 125 kg for a child above 10 (p. 124, conversion based on 27.56 kg/bu). Myers (1970, p. 206) records the necessary amount of grain per person in Shandong/Hebei at 225 kg/year. On a per capita basis Myers reports output per capita of 302.5 kg in Shandong and 433.5 kg in Hebei, suggesting marketable surpluses of between 34.4% and 92.7% of individual requirements.

Buck's nutritional chapter in *Chinese Farm Economy* was perhaps one of the first to provide a detailed examination of food and nutrition in rural China. For 150 farms in Chihli province in 1922 (p. 368), about 74.4% of food was produced on the farm, including between 91.6% (kaoliang) and 100% (wheat) of grains and 87.8% of soybeans. Major expenditures were reserved for fruits and vegetables. A total of 84.2% of calories came from grains, 14.2% from legumes, with only 2.4% coming from fruits and vegetables and only 2/10th of 1% from animal products. In a 1923 study of 144 farms in Honan (p. 369) between 77.8% (kaoliang) and 100% (wheats and barley) of foodstuffs were grown on the farm, contributing to 80.3% of total calories. Another 11.2% of calories were obtained from sweet potatoes, suggesting that only 8.5% of calories were derived from other food stuffs including legumes, and only 6/10th of 1% from meat products. Similar results were found for 149 farm households in Honan in 1923 and 217 farm households in Jiangsu in 1924. This latter survey is remarkable because it indicated that while virtually all cereal grains for consumption were grown on farm, 45.2% of rice was purchased off farm, with rice contributing to 75.1% of total calories and 60.1% of protein (p. 371).

The nutrition study in *Land Utilization* was prepared by Wen-yuh Swen from Nanjing and Leonard A. Maynard from Cornell University. Their results and observations are in line with Buck's earlier study, but went further in terms of measuring surpluses and deficiencies in calories, protein, and other nutrients. Food consumption and the nutritional value of that food was determined largely by location. In the North China Plain, for example, 80% of calories came from food produced on the farm, while in the rice regions only 25% of surveyed households obtained 80% of calories from home production. Maynard and Swen note that "*in a country where farms are so small and labor is so plentiful, it is rather surprising to find only 61% of the consumption of vegetables other than potatoes from the farm and the remainder purchased. The explanation in part, at least, is that vegetable growing is considered a specialized occupation of vegetable gardeners near cities, and the farmer usually does not even know how to grow many of the common vegetables*" (Buck 1937c, p. 402).

On measuring energy, Maynard and Swen used a 2800 calorie/day measure which was lower than the 3000 calorie/day measure used in the West. Based on this measure they found that, on average, food energy consumption was above the standard. There were, however, regional differences. The caloric intake in the rice region was 7% higher than in the wheat region. However, in the wheat regions around 40% of households consumed less than 2800 calories/day, while only 17% of households in the rice regions were deficient. Even within regions the variations were great. In Buck's winter wheat–kaoliang area the average intake was 20% above the standard, but in 11 of 33 counties surveyed, the intake was below, with one district of 1406 calories nearly half. Across the 136 districts studied, six consumed calories/day in excess of 5000 while 29 or 21.8% had calories below (Buck 1937c, p. 409).

Unfortunately, the nutritional data from *Land Utilization in China* was not included in the rediscovered data. While averages are indicative of general conditions it is ultimately the distribution that matters. Aggregated variance across all regions appears to be high. The average intake was 3282 calories/day with a standard deviation of 761. Using these values and assuming a normal distribution, the chance likelihood that any district would consume less than 2800 calories/day is 26.3%. If we use the 3282 as the most likely and apply the reported minimum of 1406 and maximum of 5000, the bounded PERT distribution indicates that 31% of districts would fall below the standard. Since the variance measures are smoothed the whole picture cannot readily be measured, but it is fair to say that the variance within each sampled village or county would be much higher. For example, a village with calories/day in the neighborhood of 2800, when averaged across 100 households, would have nearly 50% of the households below the standard, yet the village on the whole would be recorded as being above the standard.

This food economy raises some very specialized issues that cannot so easily be measured. As Fei reported, the household would calculate how much grain would need to be stored to support the consumption for all household members through to the next harvest. Any surplus above this amount could be sold and the proceeds used to purchase vegetables, fruit, and other foodstuffs not grown on the farm. On a safety-first principle the surplus would vary with crop yields, with more food and other household consumables purchased in good years, but less in low harvest years. Buck's (1937b) measures of normal yields to best yields is 87.8%, with a standard deviation of 6.6%, so in most areas yields were reasonable over the study period, yet a large number of households were still food deficient. Also, it is unclear as to what impacts the various calamities and conflicts across the 1929 to 1933 study period, as summarized in Chap. 3, might have had on consumption.

Zhou et al. (2018) address some of these issues. Food insecurity generally and the presence of "nutritional poverty traps" (NPTs) more specifically are determined by the endogenous interaction of nutrition, wages, and production. An NPT exists when food-insecure households experience an agricultural productivity shock, causing farm wages to decrease and forcing households to reduce more expensive, micronutrient-rich foods in order to maintain caloric intake from staple foods to avoid hunger in the short term. In turn, affected workers are physically incapable of doing hard manual labor. Low labor productivity implies low wages and low levels of nutrition. If the NPT exists, then the following hypotheses should be verified: Hypothesis 1: The productivity has a positive impact on wage; Hypothesis 2: The wage has a positive impact on nutrition; and Hypothesis 3: The nutrition has a positive impact on productivity. The idea is that greater productivity will lead to increases in wages, so that not only is there abundant supply of nutrition from farm productivity, but higher wages will enable greater food purchases.[21] Zhou et al. (2018) test these relationships using Buck's data from the statistical volume. They estimate a simultaneous three-equation nutrition–wage–production model using three-stage least squares (3SLS) and a number of instrumental variables to capture unobserved endogenous relationships. Additional variables include binary instruments for drought and flooding in the year of survey and previous year, as well as instruments for communist activities; military conflict with communists, bandits, and warlords; and the Northern War between the Shaanxi and Shanxi warlord governors and the national government in Nanjing.

Whether or not an NPT existed in the 1930s is difficult to assess. Zhou et al. (2018) provide some interesting insights into the agricultural economy of the 1930s. They provide two models. The unrestricted model includes control variables for calamities and conflicts, whereas the restricted model does not. Both models are sign consistent on the variables of interest. First, there is a negative relationship between wages and productivity which could signify a reduction in effort in farm work as familial labor seeks higher wage off-farm employment.

Second, there is a positive relationship between calories and productivity, suggesting a direct link between food consumption and the energy required to cultivate.

Third, there is a negative relationship between productivity and the wage rate, which is significant for the unrestricted model only. At first glance, this is a counterintuitive result since one would expect that more productive farms would retain their own labor, reducing labor supply and

[21] As productivity increases the marginal productivity of familial labor increases, which reduces the numbers entering into the wage market. As supply of labor falls, wages rise.

increasing wages. But this is not what the wage variable measures. It measures the wages collected by the household. In the unrestricted model the relationship is negative but not significantly different from zero. Instead, it appears that the presence of a drought in the previous year had some influence. This had a positive relationship, suggesting that a previous drought led to an increase in wage rates, perhaps because of the shortage of labor due to land rehabilitation. Combined, the results suggest that while higher wages lead to a causal decrease in farm productivity the opposite result does not hold, so that an increase in productivity has no causal effect on wages.

Fourth, there is a statistically causal positive relationship between nutrition and wages, and also a significant positive effect on caloric intake from wages. This suggests that workers with higher energy are more productive and can command a higher wage, and symmetrically those households receiving a wage will consume more calories. This result also suggests that households that do not have access to wage markets have a lower caloric intake, which would support the argument for involution.

Fifth, there is a positive relationship between productivity and caloric intake, but this is significant only for the unrestricted model. The fact that this is not significant when calamities and conflicts are considered suggests, generally, that there is no relationship between production and nutrition. Here Fei's observations, as mentioned, become important. Fei, and even Huang, provided specific measures of how much food is stored every year. The safety-first farmer will set aside a fixed amount of grain in amounts designed to cover requirements until the following harvest. Critically, this does not change from year to year unless the makeup of the household changes. Thus, even in good years with healthy crop yields, the household will store the same amount of grain as in previous years. With the exception of some modest purchases on vegetables and fruit, and some meat products, diet remains constant. To bolster this argument there was no difference in caloric intake in years where droughts are indicated, but there was a statistically significant reduction in calories in the year following a drought or flood.

As with equilibrium and poverty traps, NPTs are difficult to verify. In all of these situations the word "trap" suggests something other than transitory. Nonetheless, the results in Zhou et al. (2018) support the conditions required for an NTP to be present, while anecdotal evidence points to the potential presence of a NTP for large segments of China's agricultural economy. If correct, this places the economy near, or at, subsistence levels.

1.10 Agricultural Credit

Formal facilities for agricultural credit in the 1930s were underdeveloped. Chapter 12 examines this in greater econometric detail, but by and large the bulk of credit relationships recorded by Buck were from informal sources. Throughout China, a popular form of credit was through money-loan societies (*qian hui*) that operated very much like rotating savings and credit associations (ROSCA). These were comprised of a number of members who would donate cash to the instigator, and over time receive a share of the group's periodic contribution plus interest. Other forms of credit were between friends and relatives, sometimes with interest and sometimes without. Myers reports from 1936 surveys in Hebei and Shandong that nearly two in five and one in four households respectively spent more than they earned, requiring some form of borrowing to make up the difference. About 48% of Hebei's rural residents were in debt, as were 28% of those in Shandong. Borrowing was recorded for weddings and funerals as well as livestock purchases and grain.

The formalization of agricultural credit first started around 1915 with the formation of the agricultural and industrial banks. These banks were designed around the German Raiffeisen system and were to assist in the formation of, and make loans to, agricultural credit cooperatives and societies. These attempts had fizzled out by the early 1920s, but the idea behind the Raiffeisen system did not. In 1921, the China International Famine Relief Commission (CIFRC) undertook a study on agricultural credit. Ultimately, by 1924, efforts at credit cooperation were taking hold and pushed by John B. Tayler, a Briton who headed the Economics Department at Yanjing University in Peking. Tayler served on the Credit Committee of the CIFRC as did Buck. Buck's colleague Paul Hsu, supported by Buck and Tayler, started the first credit cooperative/society in 1924. While these were private efforts driven by academics, the period of rural reconstruction, starting in earnest after the Northern Expedition in 1928, started to see the rapid development of credit cooperatives. By 1929, the first year of Buck's study, there were 818 credit societies, with 21,934 members and loans granted of $122,414 (silver). By 1933, the end year of Buck's study, there were 5335 cooperatives, of which 4187 were credit cooperatives. These numbers continued to grow even during the Sino-Japanese War, peaking at around 75,883 in 1944 (Fu and Turvey 2018, Chapter 14, pp. 374–376). Although never fully implemented, the nationalist government in 1929 did prepare a "blueprint" to develop and modernize rural financial institutions. Ultimately these efforts led to the formation of the Farmers Bank of China and other progressive steps in rural infrastructure and financial development under a program of Nong Ben Ju (Farm Credit Bureau) (Fu and Turvey 2018, Chapters 12 and 13).

Of course, these efforts at developing credit markets had no meaningful bearing on the credit conditions faced by respondents in Buck's survey; their predominant source of credit was familial.

1.11 Summary

China's agriculture in the 1930s was rudimentary and largely impoverished. Technologically backward, the country relied significantly on human labor and working animals. Mechanization was largely unheard of. The remainder of the chapters in this volume address various aspects of Buck's microdata. These include investigations into yield distributions, productivity and economies of scale, land tenure, economic inequality, and agricultural credit. The granularity of data available allows assessments that could not previously be made. And this assists in the broader understanding of where in the spectrum of economic growth China had settled, and under what economic conditions. As for some of the broader subjects discussed in this chapter relating to equilibrium and poverty traps, the chapters contribute in their own ways to an understanding of the situational complexities involved. Buck's survey fell between the warlord era and the Northern Expedition to unify China, and the beginning of Japanese hostilities that led to the Sino-Japanese War. But within the period of survey there were enough calamities and conflicts to raise concerns about the reality of economic conditions on the ground. Nevertheless, the various chapters provide insights that are new and that provide levels of depth not previously available.

References

Buck, J.L. 1930. *Chinese Farm Economy: A Study of 2866 Farms in Seventeen Localities and Seven Provinces in China*. Chicago: University of Chicago Press.

———. 1937a. *Land Utilization in China: A Study of 16,786 Farms I 168 Localities, and 38,256 Farm Families in Twenty-Two Provinces in China, 1929–1933*. University of Nanking (1937), University of Chicago Press, Chicago (1937), and The Council on Economic and Cultural Affairs, NY NT, 1956.

———. 1937b. *Land Utilization in China-Atlas*. Chicago: University of Nanking (1937), University of Chicago Press.
———. 1937c. *Land Utilization in China-Statistics*. Chicago: University of Nanking (1937), University of Chicago Press.
Buck, J.L., and W.Y. Sever. 1925. *An Economic and Social Survey of 150 Farms: Yenshan County, Chihli Province, China (No. 13)*. Nanking University, College of Agriculture and Forestry.
Carter, M.R., and C.B. Barrett. 2006. The Economics of Poverty Traps and Persistent Poverty: An Asset-Based Approach. *The Journal of Development Studies* 42 (2): 178–199.
DuHalde, P. 1742. *The General History of China; Containing a Geographical, Historical, Chronological, Political and Physical Description of the Empire of China, Chinese Tartary and Tibet*. Vol. 1. 3rd ed. London: J. Watts.
———. 1772. *The Chinese Traveler Containing a Geographical, Commercial, and Political History of China*. Printed for E. and C. Dilly, London.
Elvin, M. 1972. The High-Level Equilibrium Trap: The Causes of the Decline of Inventions in the Traditional Chinese Textile Industries. In *Economic Organization in Chinese Society*, ed. W.E. Willmott. Stanford, CA: Stanford University Press.
———. 1973. *The Pattern of the Chinese Past: A Social and Economic Interpretation*. Palo Alto, CA: Stanford University Press.
———. 1984. Why China Failed to Create an Endogenous Industrial Capitalism. *Theory and Society* 13 (3): 379–391.
———. 1996. *Another History: Essays on China from a European Perspective*. Broadway, NSW: The University of Sydney.
Fei, Hsiao-Tung. 1939. *Peasant Life in China*. London: G. Routledge and Sons, Ltd.
Fu, H., and C.G. Turvey. 2018. *The Evolution of Agricultural Credit during China's Republican Era, 1912–1949*. Cham: Palgrave Macmillan.
Geertz, C. 1963. *Agricultural Involution: The Process of Ecological Change in Indonesia*. University of California Press, Berkeley, CA.
Goldenweiser, A. 1936. Loose Ends of a Theory on the Individual Pattern and Involution in Primitive Society. In *Essays in Anthropology, Presented to A.L. Kroeber*, ed. R. Lowie, 99–104. Berkeley: University of California Press.
Hagen, E.E. 1959. Population and Economic Growth. *The American Economic Review* 49 (3): 310–327.
Huang, P. 1985. *The Peasant Economy and Social Change in North China*. Stanford, CA: Stanford University Press.
Lee, F.C.H., and T. Chin. 1929. *Village Families in the Vicinity of Peiping*. China Foundation, Social Research Department.
Lin, J.Y. 1995. The Needham Puzzle: Why the Industrial Revolution Did Not Originate in China. *Economic Development and Cultural Change* 43 (2): 269–292.

———. 2008. The Needham Puzzle, the Weber Question, and China's miracle: Long-Term Performance Since the Sung Dynasty. *China Economic Journal* 1 (1): 63–95.

Malthus, T.R. 1888. *An Essay on the Principle of Population: Or, A View of Its Past and Present Effects on Human Happiness*. London: Reeves & Turner.

Myers, Ramon H. 1970. *The Chinese Peasant Economy: Agricultural Development in Hebei and Shandong 1890–1949*. Cambridge: Harvard University Press.

Nelson, R.R. 1956. A Theory of the Low-Level Equilibrium Trap in Underdeveloped Economies. *The American Economic Review* 46 (5): 894–908.

Nurkse, R. 1952. Some International Aspects of the Problem of Economic Development. *The American Economic Review* 42 (2): 571–583.

Sinha, R.P. 1973. Competing Ideology and Agricultural Strategy: Current Agricultural Development in India and China Compared with Meiji Strategy. *World Development* 1 (6): 11–30.

Tang, A.M. 1979. China's Agricultural Legacy. *Economic Development and Cultural Change* 28 (1): 1–22.

Tawney, R.H. 1932. *Land and Labour in China*. London: George Allen and Unwin Limited.

Winchester, Simon. 2008. *The Man Who Loved China*. New York, NY: Harper.

Zhou, Li, Jie Sun, and Calum G. Turvey. 2018. *Conflicts, Calamities and Nutritional Poverty Traps in a Peasant Economy: Evidence from Rural China, 1929–1933*. Unpublished Mimeo, Nanjing Agricultural University.

CHAPTER 2

John Lossing Buck and *Land Utilization in China*

Calum G. Turvey

2.1 Introduction

John Lossing Buck was born on November 27, 1890, and grew up on a farm on Freedom Plains Road, outside of Poughkeepsie in Dutchess County, New York. His grandfather and namesake, John Lossing Buck, was one of the first settlers in the town of LaGrange, where he farmed until his death on January 2, 1918. A farm boy, Buck always took an interest in the science and economics of agriculture and at one time, probably during his high-school years in Poughkeepsie, he experimented with alfalfa production. Graduating in 1910, at the age of 20, Buck enrolled that year in the College of Agriculture at Cornell University, where he was heavily influenced by the farm management studies of George F. Warren and the practices in all matters of agriculture, including the studies of rural life and sociology then being promoted by Liberty Hyde Bailey. Working his way through university as a manager to a rooming and boarding house and head waiter in a restaurant, Buck's interest in scientific agriculture deepened. Warren at that time was heavily

C. G. Turvey (✉)
Cornell University, Ithaca, NY, USA
e-mail: cgt6@cornell.edu

H. Hu et al. (eds.), *Chinese Agriculture in the 1930s*,
https://doi.org/10.1007/978-3-030-12688-9_2

involved in gathering production data, income assets, and other financial and production information pertaining to farms using the survey method. Farm management and production economics were at that time infant disciplines and Warren used the surveys to cross-tabulate economic relationships to determine which factors best expressed farm profitability, labor income, returns to assets, management, equity, and the like. Overarching these new disciplines was a deep-rooted interest in rural life. Meanwhile, in 1908, Liberty Hyde Bailey, at the behest of President Roosevelt, started work on the Rural Life Commission. While Buck was at Cornell the results of these efforts were being published and widely discussed. In addition, both Warren and Bailey were involved in certain aspects of rural credit and were building credit models around cooperatives based upon the German model.

Upon graduation from Cornell, Buck took a job as a farm instructor for juvenile delinquents at the New Hampton Reformatory in New Hampton, New York, but finding that work unsatisfactory and uninspiring looked elsewhere. Raised as an evangelical Christian, Buck considered missionary work and enquired of the Presbyterian Board of Missions whether there was any work for an agriculturalist. He was first offered a position in India but he held out for a position in China, which came in late 1915 with a request from the Madison Avenue Presbyterian Church in Nanhsuchow, Anhui province.[1,2] After a 29-day trip on the S.S. *Nippon* he arrived in Shanghai in late November 1915, and almost immediately enrolled in the Nanjing Chinese Language School where he remained until June 1916.

[1] Pugh, J. (1973) Coleman, G.P. (1962) Interviews with J.L. Buck. Cornell University Archives, Kroch Library, September 27.

[2] In a letter (undated) from Pugh to Buck, circa June 1974, Pugh writes "I trust that you will remember the nice interview you gave more than a year ago. As I soon thereafter came across a good deal of material concerning your work in China, my paper came to center around your work." This indicates that Pugh was originally interviewing for perhaps a much broader view of agricultural missionary work in China, but after meeting Buck decided that focusing on Buck provided enough material. In a letter from Buck to Pugh, dated July 1, 1974, Buck approvingly responds, "Thank you very much for your write-up of my work and thanks for doing such a good job." Buck also points out three slight errors that were never corrected in Pugh's thesis. These are page 4, change Stanley (Warren) to George F. (Warren); page 4 change Henry Coffin to George C. Hood; and page 64 (third line from bottom) add, after Stanley Warren, "son of Professor George F. Warren."

2.2 Nanhsuchow

By all accounts Nanhsuchow was a desolate place, situated about 210 miles north of Nanjing. Together, Nanjing, Hwai-Yuen, and Nanhsuchow made up the area covered by the Kiang-An Mission. The mission was set up in Nanjing in 1876 and was home in 1916 to many missionaries and affiliated missionaries, including Joseph Bailie and John H. Reisner, who would increasingly play important roles in John Buck's journey. The University of Nanjing had programs on teaching, medicine, agriculture, and forestry that Buck would soon take advantage of. Enrollees in agriculture courses in 1916 numbered 13 and in forestry 39.

Buck himself found little to admire about Nanhsuchow. In letters home he described the place as one of the most impoverished in China: a desolate town comprised of one-story earthen houses huddled within a 10 foot wall. The countryside of north Anhui was a flat and deforested plain, subject to blistering heat, frequent winds, and faminous floods. And with factions rising against the declared emperorship of Yuan Shih-k'ai, the region was politically tense.[3] Nonetheless, Buck was optimistic about what lay ahead. He described the Chinese as a "queer people" who could learn much from Western agronomists, both with regard to theory and practice: "*I am mighty glad I have come here at least so far. There is plenty of work here. The field is unlimited, almost infinite.*"[4] While studying the language, Buck started to collect data on agricultural conditions around Nanjing. He spent two months in the fall of 1917 in Szi Djou, where he was the only foreigner, and this helped his language studies tremendously. Whatever the barriers that might have existed between West and East, Buck appeared to have made a good impression upon both Nanjingese residents, who provided an acre of land, and Szi Djou residents, who provided five more, so that he could run his seed experiments. He was also provided with land in the south suburb of Nanjing Station for the same purpose.[5]

Buck understood agricultural extension from his days at Cornell, but he realized that without data of any type he was helpless to do any good. The

[3] Conn, Peter S. (1996) p. 57.

[4] Conn p. 56. Conn is referencing letters currently archived with the Nora Stirling Papers, Randolph-Macon Woman's College Archives.

[5] Reports of the Missionary and Benevolent Boards and Committees. Board of Home Missions of the Presbyterian Church in the United States of America. Presented to the General Assembly, at Columbus Ohio, May 16, 1918.

difference between being an agricultural missionary as opposed to an agricultural extension agent was not clear to Buck, and so he proceeded with his fieldwork on a scientific basis, and initiated discussions of missionary work elsewhere. For his first attempt at outreach, Buck enlisted the help of a Mr. Shao, the son of the local postmaster, and developed a three-month class for 12 resident landlords to spearhead improved agricultural practices. He also taught a high-school class in agriculture and held a short course in agriculture for the young men in the region.[6]

Buck and the farmers covered a variety of topics. With increased demand locally for eggs they discussed how to improve and enlarge the poultry industry. They talked about best practices for improving the fruit industry, and what farmers could do during the idle winter months instead of dealing with boredom by gambling. With widespread illiteracy, few farmers could read, but Buck planned a reading course on which ten farmers enrolled. These men came from different districts and tended to be large landowners, so Buck was hopeful they would relay information and practices to their tenants and other local farmers. There were enough small successes in these efforts to embolden Buck in his view that there was a place for scientific agriculture in China, and that with science and extension, the intergenerational constancy of agricultural practices could finally move forward. Buck also took an interest in river conservancy and closely followed the progress of the Hwai River Project, which was run in part by the American Red Cross in Anhui.[7] For this crop research, he tested 63 varieties of wheat seed from across China, the USA, and Japan, 26 lots of barley from across China, and 18 lots of sesame seed secured from different locations throughout China. He used a plot test to evaluate eight varieties of American wheat and two local varieties, as well as 20 varieties of beans from across China. He was able to show, for example, that the American wheat yielded as much as the best of local varieties (18 bushels/acre greater than the lowest variety) but had much stiffer straw. Cooperating with Nanjing College of Agriculture, he also tested American cotton, American corn, American sweet potatoes, spring barley, oats, alfalfa hay, and other crops. But this was a daunting task, and at times Buck saw these

[6] Buck, J.L. "Americans who have assisted in the improvements of Chinese Agriculture" mimeo. Undated. Department of Manuscripts and University Archives, Cornell University, Ithaca N.Y. Original provided to C.G. Turvey from Buck's daughter Rosalind Lewis-Smith, Buck, J.L. (1920), Buck, J.L. (1919).

[7] Buck, J.L. (1917).

efforts as hopeless, as the Nanhsuchow field encompassed three magistracies comprised of three walled county seats, 500 market towns, some 20,000 villages, a population of about 2 million, and one agricultural missionary. He was working alone, with no advanced degree, limited interaction with the specialists at Nanjing, and with funding and land allocation too small to be effective.

But it was the scientific basis of agriculture that Buck pursued, and not the immediate impact on farmers. What Buck observed at Nanhsuchow made him realize that even if he were to succeed in his research and outreach, too many poor Chinese farmers could never get ahead because they had too limited an amount of land and capital with which to work. In one way his work appeared futile given the conditions and circumstances, but, in another, still worthwhile in the longer term. Buck was pushing for more government involvement and greater public investment in scientific agriculture so that the practices, education, and outreach of this field could be extended. Observing the water conservancy project opened Buck's eyes to the political realities of the new republic, which, if not improved upon, would stymie any hope of economic growth: "*So long as needed public improvements are left to the farmer, they can never be brought about. It is the business of the provincial and central governments to make these improvements, and until this is done, the governments have failed of their whole duty.*"[8]

By 1919 Buck saw agricultural missionary work as one part extension and one part evangelical work. Buck was acutely aware that farmers enjoyed the science and lessons, but not the religion. He proposed that agricultural education should be part of all primary- and middle-school curriculums at mission schools. These schools provided a pathway to secondary education and university, and at universities, particularly Nanjing, there were programs in agriculture and forestry. A more pressing rationale was that educating young men would "*develop character and a sense of responsibility; there is no better way of doing this than by placing in sole charge of a student animals or plants for which he must care. One of the first things a boy with a garden has to learn is not to steal his schoolmate's produce. He must learn to respect the rights and properties of others. Another thing,*" Buck wrote, "*he has to learn is the taking care of his tools. The cleaning of his tools and the returning them to their proper place after he has finished using them, furnishes considerable opportunity for developing a sense of responsibility in*

[8] Buck (1917) River Conservancy in Northern Anhwei, p. 774.

the boy."[9] Buck had no doubt developed some opinions about the character of the Chinese farmer that he thought needed addressing. But he also saw that Chinese farmers lacked dignity and that dignity could be realized through better education. And with this dignity might come an appreciation of nature. Echoing perhaps the former Cornell Dean of Agriculture, Liberty Hyde Bailey, Buck claimed, "*Few people after a proper introduction to plant and animal life, fail to have a desire for closer association with growing things, and this desire is self ennobling (sic). Those who have worked with their hands in the earth, who have planted in seedtime and reaped in harvest, know something of that nearness to God, which comes from a realization of the wonders of His creation*."[10] And here was the crux of his brand of evangelism: God does not bring man closer to the land, it is the land that brings man closer to God. What Buck sought in 1919 was a national strategy for agricultural extension that would permeate all aspects of rural teaching at mission schools. Cooperation, centralization, and standardization should be a goal of agricultural missionary work.

At the same time, the College of Agriculture and Forestry at the University of Nanjing was turning out a few good Christian men in the fields of agriculture and forestry. When John Reisner offered Buck a position at the University of Nanjing he quickly accepted. Reisner argued that Buck could do far more for Chinese agriculture as a professor, and Buck agreed.[11]

Between his arrival and June 1916 Buck spent much time learning both the language and the lay of the land, but in the heat of that summer he found his way to the cooler mountainous region of Kuling in Jiangxi province, where he met Pearl Sydenstricker. What attracted Pearl to Buck is not really known, since in her later biographies Pearl S. Buck speaks little of him. A point of introduction might have been that Buck had attended classes at Cornell with the iconoclastic and radical Chinese author and scholar Hu Shi, who was a favorite of hers. Other than that, she was likely impressed by his high moral tone, enthusiasm for enriching the lives of Chinese farmers, and his technical knowledge of agriculture. Although she and Buck had been alone together no more than five times, by January

[9] Buck (1919) Agriculture in middle schools, p. 309.

[10] Buck (1919) Agriculture in middle schools, p. 310.

[11] Henry S. Coffin to Cleland B. McAfee, December 9, 1932. Archives of the Presbyterian Board for Foreign Missions. Presbyterian Historical Library, Philadelphia (Presbyterian Archives), File 82:45–49. Cf. Pugh (1973) p. 14.

1917 they were engaged, and they were married on May 17, 1917. Their marriage was contentious, and ultimately ended in divorce, but in their early years—20 years before she received the Nobel Prize in Literature in 1937 for her novel *The Good Earth*—she recalls "*as freshly as though it were yesterday the world into which it transported me, a world as distant as the one I was living as if it had been centuries ago. It was the world of the Chinese farmer*" (p. 129). Her experiences in Anhui province provided the varied themes of warlords and the plight of Chinese women, and laid the foundations for her first book, *East Wind–West Wind* (1930). And on *The Good Earth*, Pearl Buck writes (2013, p. 250), "*My story had long been clear in my mind … it had shaped itself firmly and swiftly from the events of my life, and its energy was the anger I felt for the sake of the farmers and the common folk of China.*"[12] Indeed, the central figure of *The Good Earth*, Wang Lung, was based on an Anhui farmer introduced to her during her early married life as one of Buck's first students at the agricultural mission.

Buck was appointed to develop a program in agricultural economics at Nanjing University, but was soon appointed as interim Dean of the College of Agriculture and Forestry between October 1920 and May 1922. He taught courses in agricultural economics, farm management, rural sociology, rural organization, agricultural markets, and farm engineering. By 1921 it became clear to Buck that using American textbooks for the low-technology, often backward type of agriculture practiced was futile. To remedy this, Buck convinced the university president A.J. Bowen and registrar G.W. Sarvis to allow students to receive credit for returning to their home areas and obtaining farm management surveys for at least 100 farms. With this scheme in place, Buck used the resources of the university to initiate large-scale farm surveys.[13] Between 1922 and 1925 data had been obtained from 2866 farms in 17 localities in seven provinces. The first of these was an investigation into Anhui province, which Buck submitted to satisfy a Master of Science degree in agricultural economics under the supervision of W.I. Myers at Cornell University in 1923. Myers would ultimately serve as the first Governor of America's Farm Credit Administration in 1933, and was a student of George F. Warren and Liberty Hyde Bailey, who had been pushing for agricultural cooperatives. Myers, in 1924, pushed for "agricultural finance" to be recognized as a discipline in and of

[12] Buck, P.S. (2013), Harris, T.F. (1969).
[13] Buck, J.L. (1973).

itself, and he encouraged lenders to hire specialists with appropriate knowledge and training in agriculture.[14]

Buck also pushed for agricultural cooperatives in China. In 1922 the China International Famine Relief Commission (CIFRC) set up a commission to draft regulations for rural cooperative credit societies. The regulations, based on the principal of unlimited liability, were drafted by J.B. Tayler, a Briton who was chairman of CIFRC, as well as Dean Reisner and Buck, whom Tayler had met in 1921. Once approved, funds were released from CIFRC for loans to societies formed by personnel of the committees as well as universities, including the University of Nanjing.[15] The push from the latter came from one of Buck's students: Paul Hsu. Leveraging the CIFRC funds, Hsu began discussing rural credit with farmers and in 1923 organized the first CIFRC-sponsored credit society called the Feng-Run-Men Rural Credit Society at Nanjing, and another in Keng Village near Hwaiyuen in north Anhui in 1924. In addition, Buck, Dean Reisner, and Hsu guaranteed a commercial loan from the Shanghai Commercial and Savings Bank for 1925 and 1926 to set up yet another cooperative. Following these successes, and despite reservations about the credit trustworthiness of farmers, the bank advanced funds to new cooperative ventures well into the 1930s.[16]

These experiments had far-reaching impacts in showing how cooperative credit societies could advance competition and level the playing field for poorer farmers.[17] The following years, 1926 to 1928, were tumultuous on many levels. Since the early 1920s there had been rising resentment against foreigners, including missionaries. Buck was sympathetic to some of the arguments: "It [anti-foreign sentiment] *has its good features, probably, in that it will help us as missionaries to do away with some of the weak points in our methods of work .... It seems to me that this is a period of reevaluation of methods of work, and because a change is necessary is no reason for becoming discouraged.*"[18] Pugh (p. 38) notes that Buck's work in Chihli throughout 1926 was unimpeded by the rise of nationalism under Sun Yat-sen, but that he was increasingly concerned with the factionalism

[14] Turvey, C.G., & Slaybaugh, D.P. (2006).

[15] Fu and Turvey (2018) Chapter 7; Trescott, P.B. (1993); Nathan, A.J. (1965); Malone, C.B. (1923); Tayler, J.B. (1924); Malone, C.B. and J.B. Tayler, (1923).

[16] Buck, J.L. (1973) Op Cit.

[17] Fu and Turvey (2018) Op Cit. Chapter 7.

[18] Buck to Paul T. Hickock, November 24, 1926, Presbyterian Archives, 82:31–7, p. 2, cf. Pugh 38.

between the communists and the nationalists and saw little to choose from between the northern and southern factions. When the southern faction ultimately coalesced around Chiang Kai-shek against the northern warlords, the whole idea of warlordism raised Buck's ire: "*The whole political situation in China is one essentially of selfishness; each party or militarist is out for all of the power and wealth that can be obtained by it or him.*"[19]

On March 24, 1927, a large contingent of Kuomintang (KMT) troops comprised of nationalists and communists drove through both Shanghai and Nanjing as part of the Northern Expedition to drive out warlords and unify China under one government. Despite Buck's unease with the whole situation, the couple decided against evacuation. Pearl Buck, perhaps not as tuned to anti-foreign sentiment as Buck, appeared to be giddy over the prospect of invasion, writing to her editor, "*I fancy you would enjoy being with us these exciting days … We expect a battle here any day … have decided not to evacuate with the crowd. The decision was arrived at with some trepidation on account of the two infants, but we could not bear not to see what is going to happen.*"[20] The northern warlords, who vowed to make a stand at Nanjing, were ultimately defeated. During the battle, the Bucks (including their two children and Pearl's father) remained holed up in their house with some other refugees. As news spread that the invading forces were looting and rounding up foreigners, and that Dr. John Williams, the vice-president of Nanjing University, lay dead at the university gates, the Bucks were hidden in the home of a poor farmer woman.[21] Ultimately, the powerful weapons of US gunboats on the Yangtze River provided a means of escape. Under cover of the guns they were able to reach Nanjing University and stay there in relative security for a short while. From there, the Bucks were marched to the gunboats, transported to Shanghai and then to exile in Japan. What became known as the Nanjing Incident would turn the Bucks into war refugees for almost seven months. Among the possessions that Buck took with him were the manuscript files for *Chinese Farm Economy*.

But Buck was bitter: "*The prevailing Chinese opinion is that the Nanjing Nationalist Government is skating on thin ice and may topple any time. All are agreed that the new Nationalist government (of Chiang Kai-shek) is no*

[19] Buck to Paul T. Hickock, November 24, 1926, Presbyterian Archives, 82:31–7, p. 2, cf. Pugh 38.

[20] Pearl S. Buck to Lewis Gannett (March 20, 1927) Houghton Library Special Collections, Harvard University. Cf. Conn (1996, p. 90). Note: Conn uses notation on date March 27 [20?], 1927. Had the date been March 27 the invasion of Nanking would have already occurred.

[21] Conn (1996) Page 91 states that Dr. Williams was killed at his desk in his office.

better than the old regime so far as corruption and personal ambition is concerned. Chinese returned students, our own students, and members of our faculty say that of the taxes that are collected only a small fraction are ever appropriated to the purpose for which they are raised. Military taxes go largely to commanders, not to pay the inadequately dressed and fed soldiers. The Chinese in charge of establishing the new Farm Loan Bank said that much of the new tax of $1.20 (mex) per acre is kept by the tax collector himself and that only as small amount will actually reach the bank."[22] But as the events of 1927 faded, and the nationalists established their government in Nanjing, Buck began to change his tune: "*Since the coming of the Nationalist Government to Nanjing there has been an increased interest in the subjects of agricultural economics, farm management and rural sociology. This is evidenced by increased requests from government bureaus and other national organizations for information pertaining to agricultural economic conditions and for recommendations of graduates who have had training in the same subject. There is now a demand for a number of men who have specialized in these particular subjects,*"[23] and "*I found much more interest in what is now known termed in China 'rural reconstruction work' than has existed during any previous time. I found everyone talking and writing about rural reconstruction, and the officials and government bureaus all attempting some type of rural improvement work, either on paper or in a more concrete way.*"[24] The Agricultural Economics Department at Nanjing University, which Buck had set up, was now making its mark. In February 1928, at the Fourth Plenary Session of the Central Executive Committee, and with the support of Chiang Kai-shek, a special committee on cooperation was proposed. Later, in October 1928, the central executives of the KMT ordered all branches across China to include cooperative work in their political activities.[25]

[22] "Lossing Buck Tells of Visit to Nanking" Poughkeepsie Eagle News, Saturday, January 7, 1928. p. 3. Also Buck to friends, November 13, 1927, Presbyterian Archives, 82:33–7. p. 5. Cf. Pugh p. 42 in part.

[23] Buck, J.L. (1927) Personal report of April 1, 1928 to March 31, 1929. Presbyterian Archives. 82:32–18, p. 3. cf. Pugh p. 40–41.

[24] Buck, J.L. (1933) Personal report of April 1, 1932 to March 31, 1933. Presbyterian Archives. 82:46–11, pp. 1–2. cf. Pugh p. 61.

[25] Fu and Turvey (2018) Op Cit. Chapter 8.

2.3 Chinese Farm Economy

Throughout 1928 and 1929 Buck continued and completed his work on his first book *Chinese Farm Economy*.[26] F.A. Pearson, reviewing the book for the *American Economic Review* in 1932, wrote: "*This monumental piece of work presents the first realistic picture of the agricultural activities of that vast, unknown expanse about which so much is written and so little known ... Mr. Buck has cleared away much of the fog that has prevented us from seeing clearly the problems confronting Chinese agriculture*."[27] This book was also used to satisfy the dissertation requirements of Buck's PhD, which was awarded in agricultural economics from Cornell University in 1933. Since 1921 Buck had dispatched his students and paid adjudicators throughout a number of provinces in China to undertake survey work along the lines of the farm management studies that George F. Warren had pioneered at Cornell. When completed, Buck had data on farm management and social conditions for 2866 farms in 17 localities across seven provinces (Anhui, Chihli, Henan, Shanxi, Zhejiang, Fujian, and Jiangsu). The land covered amounted to 21,000 acres (or 66,000 mou), with a population of some 17,000 persons, and the work had required a capital investment of $5 million in Chinese currency (silver dollars). The book investigated for the first time issues relating to farm layout and land utilization, farm size, capital investment and profitability, economies of size and scale, ownership and tenancy, crops and cropping systems, livestock and fertility maintenance, farm labor, the farm family and population, food consumption, and standards of living. Buck was pragmatic about the choice of location. In most cases, villages were selected because the student surveyor lived there. This has raised some objections because of the possible selection bias of students' friends and relatives or more educated farmers. But from Buck's point of view any loss of accuracy in terms of the accuracy of the bookkeeping or potential biases in location selection were offset by the accumulation of data that could correctly interpret the essential characteristics of the farm business.[28] On the other hand, by selecting

[26] Buck, J.L. (1930) "Chinese Farm Economy: A Study of 2866 Farms in Seventeen Localities and Seven Provinces in China".

[27] Pearson, F.A. (1932).

[28] A more critical view or this work can be found in Randall E. Stross' (1986) study of American agriculturalists in China with the chapter entitled "Myopia" being a biopic on Buck. Stross takes exception to the methods employed in the survey and data gathering, and more generally Buck's deployment of American farm management and agricultural economic principles that may not hold up under scrutiny given the tumult of the Republican era.

surveyors familiar to villagers, a level of trust was established that would allay fears that the data would be used to implement taxes or lay the groundwork for the seizure of land by militarists. In addition, as the survey instruments were lengthy and detailed, using local individuals ensured an avoidance of the tiresome urge of respondent farmers to walk away in the middle of the survey. Additionally, concerns have been raised that because the surveys took place over multiple years nominal prices and yields might differ by year and location without proper adjustment.

Nonetheless, what Buck provided was the first comprehensive study of China's agricultural economy. From the opening pages on farm layout he notes that on average farms are comprised of 8.5 separate plots and that the average distance between plots is 0.63 kilometers. He notes that the distance in North China is nearly 2.6 times that of East Central China, and then links this as a possible explanation as to why carts are more prevalent in the former than the latter. He quantifies in the third chapter the distribution of farm sizes, which averaged between four and five acres. Working this land was about 2.29 man-equivalents, generating sales of about $376 silver dollars per farm, with net earnings averaging $278 after imputing a cost for unpaid family labor, but with these incomes being twice as large in the East Central area when compared with North China. He examined the relationships between labor and capital, labor and earnings, and farm size and profits, painting a distribution of economic efficiencies that had hitherto been unknown and unmeasured. He identified various rental systems and attempted to determine the beneficiary in terms of return to capital. Buck also determined that rents were excessive and did not adjust for the elements of tenant management that benefited the landlord based on the opportunity cost principle. He examined what crops were grown and in what seasons and in what proportions, recorded yields and measured productivity, and did likewise for farm animals with regard to labor and markets.

Shortly after the publication of *Chinese Farm Economy*, Buck returned to Ithaca, New York, to complete the requirements for his PhD under the

Buck's focus on efficiency, particularly with respect to tenancy, ignored the greater dynamic of agriculture in China. I suppose we are somewhere in the middle. Our reading of Buck's focus on tenancy is in part due the communist claims of inequality and abuse as a hypothesis to be tested. As discussed in Chap. 4 this aspect of his research was vilified by the communists in the early 1950s. As for Stross' critique that Buck ignored the conditions and politics of the day, we agree, and Chap. 3 provides an extensive review of the various calamities and conflicts that China faced during the survey period.

supervision of Stanley Warren. As mentioned, he used his book as the basis for his dissertation work and also took the opportunity to expand his economics base to macroeconomics and the gold standard. But even before *Chinese Farm Economy* was published, Buck had initiated a program of study to investigate land utilization that would ultimately dwarf this work.

2.4 LAND UTILIZATION IN CHINA

Perhaps the most ambitious survey undertaken anywhere at that time, was Buck's attempt to survey land utilization across China.[29] It took place from about the middle of 1929 through to the middle of 1933, and it took another three years after that to complete the works in three volumes: *Land Utilization in China: A Study of 16,786 Farms in 168 Localities, and 38,256 Farm Families in Twenty-Two Provinces in China, 1929–1933*, *Land Utilization in China: Statistics*, and *Land Utilization in China: Atlas*. The main book summarized and interpreted the actual statistics, which were, for the most part, presented in aggregated form. The statistical volume was provided without any accompanying text and provided all data points summarized at the more granular village level. The *Atlas* volume included a comprehensive survey of rural populations and vital statistics and is not generally considered as part of the land utilization study. This work as a whole was considered so monumental that in 1938 Buck was awarded one of the highest civilian awards by Chiang Kai-shek: the Order of Brilliant Jade with White Cravat and Red and Blue Borders.[30]

The idea for a study on land utilization originated with the United States Department of Agriculture (USDA) and the Institute for Pacific Relations in 1927. By 1928, when J.B. Condliffe and L.T. Chen from the institute visited the University of Nanjing, Buck was already showing results from implementing field surveys, the College of Agriculture and Forestry was pushing on a number of scientific and economic fronts for agricultural research and extension, and the National government was showing a certain confidence in the university's studies as they pressed forward with rural reconstruction. On Condliffe and Chen's recommendation, John Buck received the grant, which was funded by the Rockefeller Foundation, on behalf of the Department of Agricultural Economics in 1928.

[29] Buck, J.L. (1937a, b).
[30] The National Cyclopedia of American Biography, (1980).

Buck's proposal had three main objectives. The first was to train students in methods of research in land utilization, the second was to make available knowledge of China's agricultural policies, and the third was to make available to people outside of China, and with an interest in China's welfare, some elemental information about China's land utilization, nutrition, and population. Writing in 1933, Buck stated: "*The results of this study will make it possible to formulate national agricultural policies. Types of farming will be located and this will help determine the proper location of agricultural improvement stations, the possibility of introducing crops and cropping systems from one region to another, and will be of value in comparing different agricultural regions. The information on yields when correlated with that on soils and climate will determine the extent to which yields may be increased. The information on population will make possible the compilation of such rates as birth, death, and marriage. Population growth can thus be computed and interpreted in relation to* per capita *production and the possible standard of living.*"[31]

Unlike *Chinese Farm Economy*, which covers only seven provinces, this new study was to survey 22 provinces. Ultimately 168 localities were surveyed over the next four years. In each of these villages 100 farm households were selected for the farm survey, 20 families for the food/nutrition survey, and another 250 for the population survey. As with *Chinese Farm Economy*, the selection of locations was not purely stratified or random. Where feasible, adjudicators were asked to select farm households consecutively along streets or in particular sections of the village. This avoided cherry-picking, but it was not purely random. Likewise, villages could not always be stratified as would have been desirable. Local dialects, for example, would be prohibitive if no local person could be found to conduct the training. Moreover, areas such as southern Jiangsu, which was not under the control of the National government, could not be surveyed. Indeed, as recorded in Chap. 3, much of China between 1929 and 1933 was in the throes of one disaster or another. As the study period started, drought and famine affected Shaanxi, Shanxi and Gansu. Violence was prevalent throughout many parts of China: "*Civil war, bandits and communists have interfered with the efficiency of the work and to some extent with the quality. However, by conducting studies in the least disturbed areas and by*

[31] Buck, J.L. (1933) Personal report of April 1, 1932 to March 31, 1933. Presbyterian Archives. 82:46–11, pp. 1–2. cf. Pugh p. 67.

using only personal contacts with farmers, satisfactory information has been and is being collected."[32] An example was relayed to Cornell biographer, G.P. Coleman, about how one investigator working near a communist-held area had to run for his life when the communists moved into his village. His records were left behind, and the land utilization study was unable to record data for this location.[33] Furthermore, in 1931 China faced the worst floods in recent history, forcing Buck to suspend the surveys and engage his workers on flood relief work instead.

On the basis of data collected, Buck then divided the study area into eight agricultural areas, with some subdivision into smaller areas defined by land utilization. (These regional definitions are used today to define China's agricultural map.) The larger divisions were referred to as the "wheat region" and "rice region" and were demarked generally by the North China Plain and farmlands south of the Yangtze River (Fig. 2.1). As shown in Fig. 2.2, the wheat region was broken down into three subregions, including "spring wheat," "spring wheat–millet," and "winter wheat–kaoliang." The rice region included five subregions including "Yangtze rice–wheat," "rice–tea," "Sichuan rice," "double-cropping rice," and "southwestern rice." Approximately 42% of households were surveyed in the wheat region and 58% in the rice region. A total of 73% of households were drawn from the winter wheat–millet (12%), winter wheat–kaoliang (22%), Yangtze rice–wheat (23%), and rice–tea (16%) areas.

The scope of study in *Land Utilization in China*, even by today's standards, is breathtaking. The book starts by examining, in broad terms, land, food and population, topography, climate, soils, use of the land for crops, livestock and fertilizer maintenance, size of farm businesses, and farm labor. Information on marketing, prices, and taxes was provided, as was that on nutrition and the standard of living. Of the 15 chapters presented, Buck wrote six of them. Other authors included John Hanson-Lowe, B. Burgoyne Chapman, Jason Thorp, Ardon B. Lewis, Frank W. Notestein and Chi-ming Chiao, Leonard A. Maynard and Wen-yuh Swen, and H. Brian Low.

[32] Buck, J.L. (1933) Personal report of April 1, 1931, to March 31, 1932. Presbyterian Archives. 82:43–5, P6. cf. Pugh p. 64.

[33] Buck to Coleman Page 22.

Fig. 2.1 The two main agricultural regions of China. Source: Buck, 1937a, p. 25

From the challenges of implementation to the statistical analysis with abacus, the study revealed a great deal about China's agricultural economy. All told, the survey investigated 16,786 farms and 38,256 farm families. This study draws from these data, and in particular the farm data from 168 villages or counties. As is discussed in Chap. 4, and elsewhere in this book, not all of the data was recovered and data was not collected from all locations.

Fig. 2.2 Agricultural areas of China. Source: Buck, 1937a, p. 27

2.5 Other Works by John L. Buck

Buck also held an extension appointment and was fairly active in agricultural extension work. One of his first efforts was in sericulture. The silk industry was in desperate need of higher quality silkworms, improved mulberry trees required for food, and, most importantly, a response to the disease "Pébrine," which could destroy up to 60% of silkworms. The Nanjing team, under the day-to-day direction of T.H. Chien and financially supported by Dean Reisner, proposed a mechanism by which every egg-laying moth was examined for the disease and, if identified, the eggs were discarded. Eventually the sericulture project would involve 67 women and 80 men to care for 2 million silkworms, who required 17 million meals per day, and to obtain nearly 585,000 layings of eggs free of the disease.

In 1928 C.M. Chiao and Buck published a paper on population growth indicating a doubling of China's population every 70 years. This was based on 4216 household surveys in Anhui, Henan, Jiangsu, and Shanxi. They concluded that "*Such increase, however, is accomplished with much human misery and with considerable economic waste because of the high birth and death rates. China's problem in addition to seeking an optimum population density is also one of producing a population at less cost and with less human anguish.*"[34]

In 1931, Buck halted survey work as the Yangtze and Hwai Rivers flooded with devastating impact, affecting some 25 million persons.[35] Charles Lindbergh had just arrived in China and, after taking a sampan into the flooded area with Buck, took to the air, providing mapping reconnaissance of the flood zone. Assisted by Cornell professor Stanley Warren, Buck prepared a survey schedule in an attempt to measure the impact from the floods, receiving funds from T.V. Soong, who headed the China National Flood Relief Commission, to survey 11,791 farm families, in 245 localities, and 89 *hsiens* (counties). They estimated losses at $45,700,000 Chinese currency, with 47.1% coming from crop losses and 23.7% from damaged buildings.

In 1932, following the "Shanghai Incident" and conflict with the Japanese, T.V. Soong asked Buck to come up with estimates of agricultural losses. In three weeks, 1483 households from 632 villages were surveyed, with findings that ultimately the losses were not as great as feared because in most areas the spring crop had yet to be planted; furthermore, despite significant bombing by the Japanese throughout the conflict zone, few villages had been seriously damaged.

While completing his doctoral requirements at Cornell University around 1932, Buck met Ardon B. Lewis. Lewis then came to Nanjing and lived there between 1933 and 1936, working on markets and prices for *Land Utilization in China*, but also on gold and silver prices. Lewis, Buck, and Lu-lwan Chang formed a government committee on the latter issue. Lewis and Chang undertook the bulk of this work, and were able to show that a strong relationship existed between the price of silver and declining

[34] Chiao, C.M. and J.L. Buck (1928).

[35] China. National Flood Relief Commission (1932).

price levels, leading to economic depression. Lewis published the results in the journal *Public Affairs* in 1935;[36] the work was significant enough to influence the government to move away from using silver and adopt a paper currency, with exchange rates tied to sterling and the United States dollar. In 1934, after the survey work on *Land Utilization in China* was completed, Buck took leave from the university, returning to the USA to work for the US Department of the Treasury as a monetary advisor on ways to correct the adverse effects on the Chinese economy of the US Silver Purchase Act of 1934.[37] In 1935 he returned to Nanjing when he was appointed Department of Treasury representative in China. He left Nanjing just weeks before the Japanese invasion in 1937.[38] In 1939, he became an advisor to the Ministry of Finance of the KMT, then in Chongqing, at times consulting with Chiang Kai-shek directly. Throughout the Sino-Japanese War he consulted with the government on the transport of wood oil to the USA to repay a loan to China.

In 1940, Buck returned to the University of Nanjing, which had by then been moved to Chengdu in Sichuan province to escape the Japanese. There he returned surveying, preparing, in 1941, a survey with Qi-Ming Qiao of the Farmers Bank of China to examine the consequences of the war on farmers over the crop year May 1940 to April 1941.[39]

2.6 After China

As the war with Japan drew to a close, Buck resigned from the university and became chief economist for the National Agricultural Engineering Corporation in Chengdu, working on developing import/export markets for agricultural implements. He returned to the USA in 1944 with his second wife, Lomay Chang, whom he had married in 1941. Chinese at

[36] Lewis, A.B. (1935).

[37] Poughkeepsie Eagle-News (1935) "J. Lossing Buck in Washington to Aid Chinese", Friday January 11, 1935, p. 1.

[38] Buck was divorced from Pearl Buck in 1935. Living alone in Nanking, Buck opened his house to Dr. Robert Wilson, the heroic surgeon of *The Rape of Nanking*, who had sent his family to the USA to escape the impending war. Wilson was soon joined by the surgeon Dr. Richard Brady, missionary James McCallum, and others, who would ultimately serve as members of the International Committee for the Nanking Safety Zone. Chang, Iris (1997) "The Rape of Nanking" Page 122.

[39] Buck, J.L., & Ch'iao, C.M. (1943). An Agricultural Survey of Szechwan Province, China". See also Fu, Hong and C. G Turvey (2018). *(Chapter 14, pp. 369–402).*

that time were excluded from emigrating to the United States under the Chinese Exclusions Act, and this might explain why Buck remained in Chengdu despite the war. It is believed that Lomay Buck was the first Chinese person to emigrate to the United States after the passing of the Magnuson Act in December 1943, which permitted up to 105 Chinese to emigrate yearly.[40] Between 1945 and 1946 Buck was a technical advisor to the Office of Foreign Relations with the USDA. He returned to China for the last time in 1946 as a member of the China–United States Agricultural Mission. In 1947 he joined the Food and Agricultural Organization (FAO) and established the Land and Water Use branch in Washington, DC. In 1951 he and Lomay moved to Rome, where they remained until 1954, when Buck became director of the John D. Rockefeller III-funded Council on Economic and Cultural Affairs in New York City.

Buck retired in 1957 but remained active in the agricultural community, and followed events in China very closely. In 1966 he published *Food and Agriculture in Communist China* with Owen L. Dawson and Yuan-li Wu for the Hoover Institute. Buck was very concerned about China's rapid population growth and the country's ability to feed herself, particularly through the "Great Leap Forward." What concerned Buck and his co-authors was the confusion in statistical reporting, and the subordination of statistics in 1958 to adhere to Communist Party goals. Buck saw these numbers as impossible, and as dangerous to China's food security if not addressed. This Buck did in the first chapter of the book, with Yuan-li Wu examining the causes of inaccuracy in Chinese agricultural statistics.[41]

John Lossing Buck passed away on September 27, 1975, at age 85.

2.7 Summary

The chapters in this book are based on John Lossing Buck's household data, as reported in *Land Utilization in China* (1937), and the rediscovered household spreadsheet data that has now been digitized. Starting off as a young agricultural missionary after his studies at Cornell, Buck traveled to China. He is credited with setting up the first program in agricultural economics in China at the University of Nanjing in 1920/1921 and from

[40] Jaminet, Marjorie K. (1944) "Dr. Buck Back in U.S. for War's Duration finds Inflation One of Great Evils Now in China" Poughkeepsie Journal, August 29, 1944 p. 5.

[41] Buck, J.L., O.L. Dawson and Yuan-li Wu (1966) "Food and Agriculture in Communist China".

this base he began to use survey techniques developed by George F. Warren at Cornell University, ultimately publishing two landmark books, *Chinese Farm Economy* and *Land Utilization in China*. By defining land areas, and measuring productivity, farm size and tenancy, labor utilization, nutrition, household characteristics, credit, and special expenditures, the work has become the standard reference point with which to make comparisons to the modern era. Aside from this legacy to agricultural study in China, the biography of John Lossing Buck is intriguing in and of itself. How does a Poughkeepsie farm boy and agricultural student from Cornell University end up as one of the first agricultural missionaries in China, then not only set up the first program in agricultural economics in China, but, after earning MS and PhD degrees, execute the most expansive set of surveys to define and describe China's agricultural economy? It is quite remarkable, even by today's standards.

References

Buck, J.L. 1917. River Conservancy in Northern Anhwei. *The Far Eastern Review*, December, pp. 772–774.

———. 1919. Practical Plans for the Introduction of Agriculture Into Our Middle and Primary Schools. *The Chinese Recorder* V50 (May): 307–321.

———. 1920. Agricultural Work of the American Presbyterian Mission and Nanhsuchow, Anhwei, China, 1919. *The Chinese Recorder* V51 (June): 412–419.

———. 1930. *Chinese Farm Economy: A Study of 2866 Farms in Seventeen Localities and Seven Provinces in China*. Chicago: University of Chicago Press.

———. 1937a. *Land Utilization in China: A Study of 16,786 Farms in 168 Localities, and 38,256 Farm Families in Twenty-Two Provinces in China, 1929–1933*. University of Nanking (1937), University of Chicago Press, Chicago (1937), and The Council on Economic and Cultural Affairs, NY NT, 1956.

———. 1937b. *Land Utilization in China-Atlas*. Chicago: University of Nanking (1937), University of Chicago Press.

———. 1937c. *Land Utilization in China-Statistics*. Chicago: University of Nanking (1937), University of Chicago Press.

———. 1973. *Development of Agricultural Economics at the University of Nanjing, China 1926–1946*. Cornell International Agricultural Development Bulletin 25, September. Cornell University.

Buck, P.S. 2013. *My Several Worlds: A Personal Record*. New York, NY: Open Road Media.

Buck, J.L., and C.M. Ch'iao. 1943. *"An Agricultural Survey of Szechwan Province", China*. Chongqing: Farmers Bank of China.
Buck, J.L., O.L. Dawson, and Yuan-li Wu. 1966. *Food and Agriculture in Communist China*. London: Hoover Institute/Fredrick A. Praeger Publishers.
Chang, Iris. 1997. *The Rape of Nanking: The Forgotten Holocaust of World War II*. New York, NY: Basic Books.
Chiao, C.M., and J.L. Buck. 1928. The Composition and Growth of Rural Population Groups in China. *Chinese Economic Journal* 1928: 219–235.
China. National Flood Relief Commission. 1932. *The 1931 Flood in China: An Economic Survey by the Department of Agricultural Economics*. College of Agriculture and Forestry, the University of Nanking, in Cooperation with the National Flood Relief Commission (No. 1). The University of Nanking, College of Agriculture and Forestry.
Conn, Peter S. 1996. *Pearl S. Buck: A Cultural Biography*. Cambridge: Cambridge University Press.
Fu, Hong, and Calum G. Turvey. 2018. *The Evolution of Agricultural Credit During China's Republican Era, 1912–1949*. London: Palgrave Macmillan.
Harris, T.F. 1969. *Pearl S. Buck: A Biography*. New York: John Day Co.
Lewis, A.B. 1935. Silver and Chinese Economic Problems. *Public Affairs* 8 (1): 48–55.
Malone, C.B. 1923. Study of Chinese Rural Economy. *The Chinese Social & Political Science Review* 7: 88.
Malone, C.B., and J.B. Tayler. 1923. *The Study of Chinese Rural Economy*. Peking: CIFRC Publication B-10.
Nathan, A.J. 1965. *The History of the China International Famine Relief Commission: East Asian Research Center*. Cambridge, MA: Harvard University, Harvard University Press.
Pearson, F.A. 1932. Review of *Chinese Farm Economy. American Economic Review* 22 (2): 295–297.
Pugh, J. 1973. *J. Lossing Buck, American Missionary: The Application of Scientific Agriculture in China, 1915–1944*. Honors Thesis, Swarthmore College, Swarthmore PA, May.
Stross, R.E. 1986. *The Stubborn Earth: American Agriculturalists on Chinese Soil, 1898–1937*, 92. Berkeley: University of California Press.
Tayler, J.B. 1924. Study of Chinese Rural Economy. *The Chinese Social & Political Science Review* 8: 230.
Trescott, P.B. 1993. John Bernard Tayler and the Development of Cooperatives in China, 1917–1945. *Annals of Public and Cooperative Economics* 64 (2): 209–226.
Turvey, C.G., and D.P. Slaybaugh. 2006. William Irving Myers: A Biography. *Agricultural Finance Review* 66: 1–6.

CHAPTER 3

Calamities and Conflict Affecting Rural China 1929–1933

Calum G. Turvey

This chapter is based on information drawn from numerous contemporaneous sources between 1929 and 1933 by C.G. Turvey. Some of this material may have appeared in verbatim form across the various chapters in H. Fu and C.G. Turvey (2018), *The Evolution of Agricultural Credit during China's Republican Era, 1912–1949*, Palgrave Macmillan, Cham.

3.1 Introduction

If one were to seek an opening to criticize Buck and his survey, it is in the contextual layering of the many calamities and conflicts that swirled around China during the survey period 1929–1933. A naïve reader of *Land Utilization in China* would pick up on little that might hint that the survey period was not normal, yet the years between 1929 and 1933 were anything but ordinary. For example, between January 1 and July 10, 1933, the National Relief Commission in China itemized the following natural catastrophes in the country's provinces: Zhejiang: floods in six districts;

C. G. Turvey (✉)
Cornell University, Ithaca, NY, USA
e-mail: cgt6@cornell.edu

H. Hu et al. (eds.), *Chinese Agriculture in the 1930s*,
https://doi.org/10.1007/978-3-030-12688-9_3

Jiangxi: floods in 14 districts; Hunan: floods in 32 districts; Henan: floods in 11 districts, drought in seven districts, frost in four districts, hailstones in 11 districts, locusts in three districts, and windstorms in six districts; Shaanxi: drought in 13 districts, frost in 31 districts, hail in seven districts, gales in 37 districts, and floods in three districts; Gansu: earthquakes in seven districts, drought in four districts, famine in 30 districts, and plague in one district; Anhui: gales in two districts, floods in four districts, and hail in two districts; Kweichow: drought in 13 districts, hail in three districts, floods in four districts, and gales in three districts; Jiangsu: floods in one district. A week after releasing this report, the Yellow River, often called "China's Sorrow," "The Ungovernable," or "Scourge of the Sons of Han," began to rise. In Sanyuan, Shaanxi province, the waters rose rapidly, drowning some 5000 farmers from both farmland and mountainous areas. Flooding spread to Henan, Hebei, and Shandong. By the time the Yellow River subsided, approximately 50,000 Chinese in total had perished, 2 million were rendered homeless, and 1 million were starving. In the Changyuan district of Hebei alone, losses were evaluated at $37,210,000 (Mexican silver), with 2223 villages flooded, 619,000 Chinese homeless, and 475,000 mou (one acre = six mou) underwater.

Meanwhile, China was an embattled country. As the Yellow River began to rise an armistice was signed by Japan and China, halting further militarism in Manchuria and areas north of the Great Wall. Between 1931 and 1932 some 222,000 Chinese, including 54,000 civilians, had been killed or wounded in conflicts with Japan in Manchuria and Shanghai. In southern China Sichuan was in the midst of a senseless civil war, with casualties running into tens of thousands. Communist forces under Zhu De and Mao Zedong were in constant battle with the Nationalist forces of Chiang Kai-shek in Henan, Hunan, Fujian, Hubei, Sichuan, Anhui, Jiangxi, and Guangdong, taking a heavy toll of the agricultural economy. In western Henan in 1932 it was reported that 50,000 people had died in the strife and in Fujian the land was laid to waste and abandoned. In June 1933 a burial detachment from the Shanghai Red Swastika (Red Cross) proceeded to Anhui to put to rest 18,000 uninterred bodies.

Even in areas that were not vested in civil war or anti-communist campaigns or communist insurgency, roving warlords left over from the post-republican period formed massive bandit gangs that laid havoc to rural regions. In some instances warlord armies reached tens of thousands and were given provincial control by the formal government. But these were not truly governing forces and exploitation of farmers ensued through dues or common confiscation of crops. To maintain the warlord army, farmers were bled dry. For example, Shaanxi province, wrecked in recent years by drought and famine, was one of the poorest in China. But poverty in China was not always driven by natural calamities. Shaanxi had been under the control of the "Christian General" Feng Yu-hsiang since 1920 and it was he who ordered the paying of a land tax throughout 1934. The nationalist government had outlawed Likin (gift in cash) in 1930 but this was simply replaced by taxes in disguise. The central government had ordered production and consumption taxes but at the local level there were also a host of taxes and duties which farmers had to pay, including land tax, poll tax, bandit-suppression duties, military dues, commissary dues, *ming-tuan* or militia dues, land registration fees, opium land duty, shares in provincial banks, provincial treasury notes, village pacification fees, rice duty, trade tax, special tax, land deed examination fees, stamp tax, and other duties and surcharges levied by local (rather than provincial) governments. Some of these taxes were extraordinarily extortionary. If the farm registration fee was not paid the land could be confiscated, the bank shares were never issued to farmers, the stamp tax was paid whether or not a shop was in the village, provincial treasury notes were apportioned and issued to farmers whether they wanted it or not, the lands planted with opium were decided by the province, but when yield was not sufficient the $10/mou duty had to be paid by those farmers who had not yet planted opium. And when the provincial government sent its agents to press farmers for these taxes, the agents had to be fully, and generously, accommodated and entertained at the expense of farmers. In the region of Hanzhong, fertile with irrigation, the net proceeds from double-cropped grain, wheat, and beans gave the farmer about $8/mou. The land tax on this was $3.50/mou, with $12/mou required to pay for other taxes and duties, meaning the farmer had to come up with an extra $7.50. In the mountainous region of Liupan the cost of production per mou was about $3 but the levies and taxes were over $11. When these could not be paid, the farmer sold or pawned whatever possessions he had (clothes, furniture, etc.) and when that did not satisfy the demand, he simply left the land, risking torture or even death for non-payment if caught.

Taxes, of course, had to be paid in cash, with no facility for credit, so, when pressed at harvest, farmers could not afford to store grain for future sale. Hence, with all farmers selling at the same time, such abundance lowered the price, so that any possible benefit of storing post-harvest was lost to taxes while the benefits of storage accrued to marketers and wholesalers. And in times of scarcity, due perhaps to drought, the farmer had no crops to sell and could not afford the steep rise in prices that accompanied famine. Even if the farmer had surplus worth selling, often the commodities could not be moved. For example, the local price of rice in Sian (Xi'an) was $20/picul, while the local price in south Shaanxi was $3/picul. A merchant could profit by transporting from south to north. But the exactions and extortions along the way by "special" tax collections, by agents and city gatekeepers, and by bandits, as well as a lack of communication with potential buyers, made the venture perilous if not unprofitable.

The drought in 1928–1930 in Gansu, Shanxi, Shaanxi, and elsewhere in China's central provinces was of varied intensity and duration, but the documentary evidence shows clear signs of a devastating famine. The main effects were noticed throughout 1929 and 1930. Buck's data reveals several important observations, the most important being that price effects differed significantly by locality, suggesting that they were perhaps more locally than spatially correlated, although no doubt there would be some systemic effects. In Shanxi province, for example, prices received by farmers rose 37% in 1929 and again by 37% in 1930 at Ningwu, falling by 60% in 1931. Similarly, in Lin prices rose by 49.8% in 1929 and another 35% in 1930, falling by 17% in 1931. In contrast, prices received by farmers increased at Tsinglo by 51.2% in 1929 and then by only 5.4% in 1930, most likely because the drought ended sooner at that locality.

3.2 Calamities and Conflict Affecting Rural China 1929–1933

If Buck's portrayal of land use in China is to be criticized it is that he never placed any of the statistical data in context. For example, farm productivity in Shaanxi in 1929 and 1930 was treated no differently and without qualification to, say, Hunan in 1933. This may have been a serious oversight. China between 1929 and 1933 was in tumult due to a variety of natural calamities and manmade conflicts and each of these in one way or another can affect the comparative value of the survey data. For example, reports in August 1930 stated that under current conditions massive amounts of land were being underutilized. Floods, insects, drought, wars, banditry, and communist disturbances were forcing farmers to leave land uncultivated.[1] Of those who remained on the land the vast majority did

[1] Shen, Chennan. August 16th, 1930. Is China facing Starvation? *China Weekly Review* 53: 422–423.

not have the land needed to adopt new technologies that could improve yields; they consumed their own stock in times of crisis, therefore depleting the supply available for general consumption. Furthermore, the calamities and conflicts on which this chapter report had much broader economic consequences. For example, while there was probably enough rice land in China to produce a crop to feed the world, there was in 1930 a shortage which encouraged hoarding behavior, driving the price up to the highest levels ever recorded ($24/200 lbs or per 'shih'). A lack of shipping and transports limited the importation of commercial or relief grain into areas depressed by famine or war, which exacerbated conditions, while in all areas the rapid rise in price reduced consumption, especially for poorer households who relied on rice as a staple. In Shanghai, dock and factory workers went on strike in order to have a "rice allowance" added to their wages.[2]

The costs of militarism on agriculture should not be ignored either. In Jiangxi alone such losses between 1926 and 1931 numbered 6600 houses burned with 22,600 casualties and property damage of $5 million; in Kweichi 3000 houses were burned with 60 casualties and $400,000 of property damage; in Yichuan 700 houses were burned with 100 casualties and $20 million of property damage; in An-fu 5000 houses were burned, with 7000 casualties reported but property damage of only $35,000. The list goes on, with maximum property damage of $40 million in Kwungfeng, 8000 houses burned in Yinshin, and 22,600 casualties in Shuichan.[3]

Yet, other reports show that certain aspects of militarism, especially those related to communist activities, advanced agricultural productivity.[4] To the farmer class many of these were attractive, especially those related to the redistribution of land and to working conditions, so it is no wonder, given conditions in the rural areas, that the communists could attract members and forces. Even so, context matters. For example, while the communists, in establishing a soviet, eliminated all forms of taxation, this was unlikely to be economically neutral for often the tax on grain was in physical commodity and not cash, and what was taxed was consumed without being necessarily sold on the cash market. Consequently, to fund

[2] Lee, B.Y. August 16th, 1930. The Rice Situation and China's Welfare. *China Weekly Review* 53: 419–420.

[3] Yen, Chia Hsi. January 3rd, 1931. What Communist Bandits have done in Kiangsi. *China Weekly Review* 55: 186–187.

[4] Chien, Yang. July 25th, 1931. Canton Rebellion Likely to Throw All China into Ranks of Communists. *China Weekly Review* 57: 297–300.

activities, systems of fees and surcharges were put in place in addition to the forced surrender of grain when needed.

For our purposes the activities related to agrarian reform were so sweeping that they cannot be ignored. On the one hand is militarism itself. This can not only destroy crops and dislocate farmers, but the commandeering of trains for troop transport meant less freight for shipment of grains and produce to markets, that contracts with middlemen could not be guaranteed for delivery, and so on. But it also meant that with thousands of farmers joining the Red Army, labor productivity would decline or wages would have to increase accordingly. In areas with large tenant populations it is unlikely that landlords would willingly invest in improvements if there were a risk that lands and titles would be confiscated and redistributed to farmers and soldiers. Credit, were it available, would melt away with declarations that all notes bearing high interest and pawn shop tickets would be considered null and void.

On the other hand, the farm economy was aided by farm reconstruction, improvement in irrigation, prevention of floods and droughts, supporting emigration to reduce farm population density, the establishment of farmers' banks and cooperative societies to provide credit on easy terms, unifying currency and weights and measures, and maintaining efficient control of waterways. In Jiangxi many such reforms were put in place, including the burning of all deeds and the removal of boundary markers to destroy all evidence of ownership. The land was redistributed to the able-bodied regardless of sex, and additional land was allocated to households that included persons with disabilities or children under age of 16 up to 25% of the normal allocation. To aid in recruitment, land held by soldiers of the Red Army would be tilled by others. The labor class, including poor and middle-poor farmers and workers, were organized into unions. The middle- and upper-class gentry were excluded from all political activities for fear they would manipulate a class struggle in order to get land back. The bandit class, including the various societies (Small Sword, Red Spears, etc.), was absorbed into the soviet and the military as long as they renounced any authority other than the soviet. In doing so the communists removed a significant element of banditry.

In Buck's study, with various villages being surveyed at different locations with varying degrees of conflict interpreting his data as if all other things were equal is likely to be incorrect. For example, except for one village surveyed in Liaoning there is no observation that might provide insights into the impact of productivity on agriculture following the

Japanese annexation of Manchuria and provinces north of the Great Wall between 1931 and 1933, nor is there any observation of villages in the battle zone around Shanghai during the Sino-Japanese conflict in January to March 1932. But not all provinces experienced conflict and not all faced a calamity. In addition, we know from Chap. 2 that Buck would not have sent his students and surveyors into areas controlled by warring bandits, warlords, communists, or nationalists for safety reasons. Further, it is not always known when exactly a particular village was surveyed. So if it was in a war zone it might have been surveyed before the war, after the war, or during the war if at some safe distance from the front lines. If the randomized survey protocol was truly practical, the *ceteris paribus* interpretation of the results could be relied upon with some sense of confidence. Because this is open to question, the results must be interpreted in the context of historical developments at the time.

3.3 Flood, Drought, Pestilence, and Famine

The most significant calamity of 1929 was the ongoing drought in China's north and central provinces, including Shaanxi, Shanxi, Henan, Gansu, Suiyuan (Inner Mongolia), Chahar (Eastern Inner Mongolia), Shandong, Hebei, and Hubei. Initial reports in December 1928 mentioned a famine arising in Shaanxi as a result of drought. The rivers Wei and Kin were drying out and there were reports of plague and locusts. A total of 91 districts were affected, with reports of young girls being sold for marriage at a price of $4 or $5 so that they would not perish from starvation. At Kuanshien in Shandong 100,000 people were reported as being destitute. The China International Famine Relief Commission (CIFRC) reported that in one *hsien* in the area between the Yellow River and Sianfu in Shaanxi, 70,000 of the total population of 120,000 were completely destitute of foods or substitutes, with the remainder living on meager diets. Dried grasses were being ground to make a sort of porridge. Draft animals had disappeared, having been eaten. In one district 100 families committed suicide rather than continue to suffer.[5] Edgar Snow provided the historical context of the unfolding disaster. Approximately 50 million persons were affected by the drought and famine in Shaanxi, Henan, and Gansu, an area that was relentlessly impacted by a host of calamities including the "quintet" of drought, famine, flood, earthquake, and locusts. The range of calamities

[5] Editorial, February 23rd, 1929. *China Weekly Review* 47: 519.

and conflicts affecting China since 1850 included the 1850–1864 Taiping Rebellion, with 20 million killed and nine provinces decimated; the 1878 famine in Shaanxi and Shanxi, with 8 million dead; the 1894 war with Japan; severe drought in Shaanxi in 1899; drought and famine in Shaanxi and Henan in 1923; an earthquake and mountain slip in Gansu, killing 200,000, also in 1923; and, in 1924–1929, continued drought and famine in Shaanxi, Henan, and Gansu.[6]

In 1929 Gansu, only in the valley of the Yellow River were crops normal. In other areas crops were poor but, if rationed properly, could have been sustainable, while in yet others the crops were devastated and, without a precautionary stockpile, famine took hold very quickly. Harry Paxton Howard wrote: "*Throughout this vast territory conditions are such as are unknown and incomprehensible in modern countries today. From eating bark of trees, grass roots, and vermin of every available kind, the starving people have gone on to eating what they hopefully call 'bread-stones,' which are simply a kind of rather solid clay with no food value whatever. Finally has come cannibalism, and the emaciated bodies of the dead are consumed to maintain, for a little while, the spark of life in the living … There is cannibalism over most of the province at present but particularly in the south. Not only the emaciated dead are eaten, but sometimes nourishing victims are found. A detachment of 10 of Col. Chao Si-ping's soldiers, near Lingshih, failed to reach a city before dark, and slept out in the open. It was their last sleep, as about all that was left of the detachment in the morning were gnawed bones and some clothing. They had been eaten.*"[7] Edgar Snow, visiting the famine regions of Suiyuan in Inner Mongolia, observed: "*stepping over a skin-draped skeleton in which a faint agitation told that life still loitered, as though waiting for an unseen hand to order it to shuffle away forever … a withered young woman who was frantically burrowing a hole into a little mound of earth, digging for the roots of a leafless (tree) that somehow had managed to escape complete dismemberment … A family group … spent a week in tramping across the dusty semi-desert of the southwest … there were four sons each with 12 mou of land, enough to support the family … to keep from starving they ate their seed grain. An opportunity came to sell their women and girls to wealthy Chinese in the east, rather than watch them*

[6] Snow, E. December 15th, 1928. China's Five Horrors. *China Weekly Review* 47: 122.

[7] Howard, H. P. June 1st, 1929. Famine and "Mohammedan" Banditry Again Devastating Kansu. *China Weekly Review* 48: 17.

starve to death, they agreed to the deal … one fellow—the ghost of a once powerful man, was particularly pitiable. The muscle had dropped away from his broad shoulders and stout arms; one could have joined the fingertips of one's two hands around his waist, and encircled his biceps with the thumb and forefinger of one hand. They all wore such rags that the greatest mystery was how the countless tatters managed to hang together as one garment … signs of famine disease. Their faces were puffed like bloated sausages, and in color their skin was like stagnant water. Their eyes, in which lingered no trace of the alert curiosity so characteristic of Chinese children, were watery and sometimes almost obliterated by the bags of mottled flesh that surrounded them."[8]

By May 1929, however, the full extent of the drought was recognized. Famine conditions in China were affecting about ten times more people than the major famine of 1920–1921 and the entirety of shipments of grain into the famine areas, ending April 1, 1929, was less than the amount of grain moved from Manchuria during each week of the 1920–1921 famine. In the famine regions of Hebei and Shandong there were no shipments of grain from the abundant stocks in Manchuria and assistance was mostly financial, with local agencies purchasing the required grain. But in other provinces where grain was not to be found the money was useless, for as grain was not being shipped it could not be purchased. Relief was not effortless. Ultimately the CIFRC leased locomotives and cars for relief purposes and were given assurance by the local military that famine shipments would travel without incident.[9]

In 1930 conditions improved in many areas but not for Shaanxi. In June 1930, the CIFRC reported a widespread drought in Shaanxi was requiring burial squads to make two rounds per day in Sian (Xi'an). On the Sanyuan Plain in Shaanxi there was only a 5% yield on crops and poor conditions were also reported in western counties.[10] In addition there were reports of a serious famine in the eastern part of Zhejiang, with the population there requiring massive amounts of food aid.[11] In the district of Paotachen, Suiyuan province, already under famine conditions, crops were damaged from a rare summer cold spell that dumped up to six inches

[8] Snow, E. August 3rd, 1929. Saving 250,000 Lives. *China Weekly Review* 48: 418–424.

[9] Clark, G. May 18th, 1929. Famine Relief in North China. *China Weekly Review* 48.

[10] Editorial. June 14th, 1930. Acute Famine Conditions in Shensi Province. *China Weekly Review* 53: 57.

[11] Editorial. June 21st, 1930. Famine Conditions in Chekiang. *China Weekly Review* 53: 110.

of snow in June onto the land.[12] In 1931 famine conditions were reported in west Henan and Gansu.[13]

An unusually warm March increased the current of the Yellow River in Tsinan, which became so strong that the river breached a dyke at Litsin Hsien, resulting in damage to 100 villages, with tens of thousands of farmers affected and sheltering on the dykes.[14] In August 1929 rains in north China caused the Yungting River to overflow, inundating ten villages and flooding 200 sq. miles of land, with many persons drowned as well as thousands of cattle and all crops lost. The floods affected nearly 500,000 people, who, in great starving numbers, were forced to live on the sides or summits of hills. Efforts at relief were hampered by the enormity of the flood and the isolation of the villages and the stranded.[15] In Shandong a 1000 ft breech in the Yellow River flooded a strip of land 50 miles long and ten miles wide, destroying the bean and grain crop in an already famine-ridden area for which the destroyed crop had been a promising aid.[16]

In June and July 1931 the *China Weekly Review* posted two short articles that were prescient. On June 13, 1931, in one of the first signs of the disastrous flooding to engulf the Yellow and Yangtze River basin, northeast Hunan was inundated with rainfall.[17] Landslides were reported and rainfall from the mountains was so intense that debris covered the valley for several square miles. Many lives were reported lost and crop losses were immense, with newly sown fields laid to waste. The second article, on July 4, 1931, discussed improved production in many areas of China.[18] Despite the drought of 1929, its impact into 1930, and various civil wars

[12] Editorial. June 28th, 1930. Summer Cold Wave Hits Suiyuan Province; Crops Damaged. *China Weekly Review* 53: 147.

[13] Editorial. April 4th, 1931. Banditry and Famine in Honan. *China Weekly Review* 56: 156; Editorial. April 4th, 1931. Hunger in Kansu Drives Population to Cannibalism. *China Weekly Review* 56: 157.

[14] Editorial. March 16th, 1929. Warm Weather Causes Breach in Yellow River Dyke. *China Weekly Review* 48: 120.

[15] Editorial. August 3rd, 1929. Ten Villages Flooded in Chihli. *China Weekly Review* 48: 446; Editorial. August 17th, 1929. North China Areas Devastated by Floods. *China Weekly Review* 48: 529.

[16] Editorial. August 31st, 1929. New Floods Reported in North China. *China Weekly Review* 50: 14.

[17] Editorial. June 13rd, 1931. Terrific Rains in Hunan. *China Weekly Review* 57: 69.

[18] Editorial. July 4th, 1931. China's Wheat Crop for 1931 Estimated at 600,000 Bushels. *China Weekly Review* 57: 195.

and strife, China was projected to enjoy the largest crop ever. North of the Great Wall in the spring wheat area, production was projected to be 146% of normal yields, with yields in Heulungkiang being 171% higher than average. Winter wheat in the southern provinces of Shanxi (49%), Hebei (74%), Shandong (105%), north Jiangsu (92%), Anhui (74%), Henan (87%), Hubei (79%), and Jiangsu (71%) were expected to be lower (86%). Although crop conditions were good in Shandong, Hebei, and Henan, fewer acres had been planted in Hebei and Henan due to poor weather conditions and warfare.

Optimism about bumper crops was soon to be dashed by the great floods of the Yellow and Yangtze Rivers. On July 11 there were reports of serious flooding across China. The West and North Rivers at Guangdong had overflowed their banks and the Yungting and Taching Rivers in Tientsin were similarly inundated. In Jiangsu the Grand Canal had risen ten feet and had submerged several villages. In eastern Henan near Tangshan, the Yellow River was overflowing and flooding large sections of farmland. Refugees were already beginning to congregate on nearby hills. Heavy rains in Nanjing saw several buildings collapse.[19] On July 18, 1931, it was reported that although conditions had become less serious in many affected places, rail traffic was disrupted by excessive flooding in Henan, with one train wrecked when the roadbed gave way.[20] In north Jiangsu one district was turned into a lake as a consequence of continuous downpour. Water on the streets was knee deep and most houses were submerged. Elsewhere in Jiangsu the dykes at Tsingkiangpu and Chuchow collapsed, flooding rice fields, with damage estimates of $6 or 7 million. Elsewhere, the Huai River had risen 20 ft and was in peril of inundating many districts in northern Anhwei if the weakened northern dykes failed. On July 7 most of Peiping (Peking/Beijing) was under water and electricity had cut out, plunging most of the city into darkness. Along the Yellow River, which had risen to 85 ft above sea level, dykes and embankments collapsed in both the upper and lower parts of the river. In Pengpu a dispatch dated July 13 stated that after ten straight days of heavy rains the city was several feet under water, with boats moving goods and people around, and that 5000 people were homeless. On August 1, widespread flooding

[19] Editorial. July 11th, 1931. Serious Floods in Many Parts of China. *China Weekly Review* 57: 242.

[20] Editorial. July 18th, 1931. Floods Cause Disaster in Many Parts of China. *China Weekly Review* 57: 276.

in Nanjing was reported, with 1000 houses having collapsed due to heavy rains.[21] Conditions in Jiangsu were so severe that by the fall the area was in a state of famine. Near Wuhu in Anhui province, the collapse of embankments flooded 200,000 mou of rice fields. Farmhouses collapsed and many livestock drowned. Wuhu was also flooded. Reports of flooding in south Shandong followed heavy rains combined with hail on July 16. Many cities along the Shanghai–Nanjing Railway were flooded. At least 30% of rice fields were damaged and granaries were washed away, requiring sampans (river boats) to be used for storage. The Chientang River in Zhejiang province was on the verge of overflowing due to heavy rains and many districts were inundated with flood waters. On August 8, 1931, it was reported that the Yangtze River near Wuhan had flooded on August 1, drowning several hundred people.[22] A dam at Hankou collapsed on August 2, and hoarding and profiteering in coal and foodstuffs was outlawed. In Anking, losses in 39 districts exceeded $4 million but, with a provincial deficit already, a ban on the export of rice was imposed. The Yellow River in Shaanxi broke its banks and flooded a wide area in Pingmin. In Tsingyuan, about 60 miles northwest of Canton (Guangzhou), 1297 persons were drowned, 10,000 were rendered homeless, 36 fish ponds ruined, 25,000 mou of land damaged, and losses totaled some $2 million. In Hunan, General Ho Chine, chairman of the provincial government, ordered all slaughterhouses to close for one day so that the wrath of Heaven might be appeased and the people spared. By August 15, 1931, the full scope of the tragedy was being realized as potentially the worst flooding since the fifteenth century, with much of central China flooded, with Hubei, Hunan, Jiangxi, Anhui, Sichuan, and Shandong hit the hardest.[23] The General Relief Committee at Nanjing reported that some 50 million persons were in distress and 16 provinces were affected by floods. They petitioned the nationalist government in Nanjing for $20 million for relief and another $5 million to be raised in Shanghai

[21] Editorial. August 1st, 1931. Central Government to Grant Relief to Flood Stricken Areas. *China Weekly Review* 57: 364.

[22] Editorial. August 8th 1931. Nanking Relief Committee Investigates Flood-stricken areas. *China Weekly Review* 57: 406.

[23] Editorial. August 15th, 1931. Fifty Million in Distress as Result of Floods. *China Weekly Review* 57: 431; Editorial. August 29th, 1931. Flood-Famine Situation Most Serious China Disaster since Fifteenth Century. *China Weekly Review* 57: 495–499; Editorial. September 5th, 1931. Further Areas Inundated in Kiangsu-Gen. Chiang Inspects Hankow Flood District. *China Weekly Review* 58: 7–13.

($25 million). As of August 11, the Grand Canal had not yet burst its banks, but the Jiangsu provincial government was alarmed by the rise of water. Starving farmers in the vicinity of Kiukiang were reduced to selling their oxen and buffalo for a paltry $4 and $5 each to buy food for themselves. Although the water was receding all hopes of a harvest were lost. A total of 220,000 persons in Hankou were rendered homeless. It was observed that farmers removed uninterred bodies from coffins in order to use the coffins to transport themselves from one place to another. Because of water levels ships could not be unloaded at Hankou. Tugboats were used for temporary storage, and food shortages in Wuhan forced the Hubei government to telegraph Jiangxi for an immediate shipment of 100,000 piculs of rice. Flood damages in Hubei, Hunan, Jiangxi, Anhui, Jiangsu, and Henan were estimated at $8 million and, in Jiangxi, some tax relief on arrears was provided to farmers.

By October 1931 the floods had subsided in most parts of central China, albeit too late in many cases for planting a winter crop. It could take years for the land to regain its former productivity, so the outlook for 1932 was not bright. Furthermore, the calamities continued into 1932 and 1933, with famine and drought unceasing in Shaanxi, more flooding in Hunan and Jiangxi, cold and drought in Gansu, flooding in Harbin of Manchuria, and drought in Anhui.

3.4 Bandits, Warlords, Civil War, and Agricultural Conditions

Banditry was rampant across China. In a letter to the editors of *China Weekly Review*, John Lossing Buck wrote of meeting farmers in a teahouse to discuss credit cooperatives.[24] A farmer said "*if any financial help is rendered to us now, it would only bring more harm than good, for our greatest distress and sorrow is no other than the bandits. We can neither work in the daytime nor rest at night … well-to-do families have taken refuge in the city. Those who are poor … hurry away and hide with their children wet and cold, in the bushes and streams of the mountainside in spite of the mosquitoes and snakes … many deaths have occurred. We do not grieve over the dead, for it is better to die of sickness than to be killed by bandits … we would rather die than suffer.*"

[24] Buck, J. L. December 15th, 1928. Letter to the Editors. *China Weekly Review* 47: 73.

Bandit forces arose for many reasons. In some instances it was the inability of the economy to absorb soldiers being released, as the national government was trying to disband its numbers after the Northern Expedition to eradicate the northern warlords between 1927 and 1928. Other bandit forces were remnant forces of former warlords, or led by warlords who had escaped the Nanjing forces.

In many rural areas remnants of the Boxer societies still persisted, along with superstitions. In Henan at least three secret societies were active.[25] The Heavenly Obedience Society was formed in 1925 by a Wang Lao-Feng, who declared that he was the "Real Dragon" and would soon be China's emperor. That organization comprised many women as well, and had both male and female military commanders under the leader. Believing in charms and spells they decided one day that that Ki Hsien would be their capital, but their attempt to claim this area saw them badly beaten, and they dispersed, with members becoming bandits or joining other societies. The Heavenly Gate Society was formed by Han Yu-ming, a stone cutter who found a stone seal and, pretending that he had cut it from a large stone, established an altar at which he referred to himself as the Old Corps Tutor. He convinced members that they were immune to gunshots and took over a city, called himself emperor, and started to collect taxes. The Cannon Society emerged in western Henan in 1927 or 1928 with the slogans "destroy the bandits" and "refuse to pay taxes." Eventually they too became bandits and had to be disbanded by the Kuominchun. Members could freely kill their enemies and burn their houses. Eventually the leader, Chang Peng-chu, was killed at Loyang. His immediate subordinates were burned alive and the remainder became bandits. The most notorious and perhaps longest lasting society was the Red Spears, which acted, depending on the circumstances, as a rural crop protection force or as bandits. On September 23, just outside of Penglai in Shandong, a large force of Red Spears was destroyed by provincial troops. These Red Spears, largely a bandit group rather than a society, had been terrorizing the area, burning 80 villages, looting and murdering even women with children who had escaped into the corn fields.

In other instances bandit gangs arose out of various prejudices that compounded over time. Buck, writing in 1928, discusses the emergence

[25] Loo, M. M. April 6th, 1929. Some Secret Organizations in Honan. *China Weekly Review* 48: 248.

of the Big Swords and Little Swords.[26] On September 10 and 11, 1928, the two groups clashed. The origins of each group resulted from migration of farmers from the north to Chinkiang in Jiangsu province, whose habits and customs differed from those of local farmers. The northern farmers, for example, built earthen houses with thatched roofs, while the southern farmers built houses of brick with tiled roofs. The northerners were mostly economic refugees and lived in poverty. The two groups rarely mixed or intermarried. When the nationalists took control in 1927 an upsurge in banditry took place, and, as it turned out, the locals were, by far, more affected than the immigrants. Consequently, the local southerners believed that the northerners were part of a bandit group (which some might have been) and attacked the immigrant villages, burning thousands of homes. The immigrants then organized "Little Sword" societies and in response the locals organized "Big Sword" societies. Both societies were similar to the Red Spears in that they believed themselves to be impervious to bullets. Indeed, one enlightened farmer refused to join the Big Swords unless he could first shoot six bullets into the leader's chest and the leader survive. The leader declined.

At first the Big Swords and Little Swords would parade around, seeking more inductees, until both societies grew to a threatening size and became well armed. In the Chinkiang conflict, which resulted from an attack on the Big Swords by the Little Swords, the Little Swords amassed some 2000 men (including some bandits) with one machine gun, 100 rifles, and 100 pistols, with the remaining force using swords, spears, and farm tools. Six villages were burned to the ground with 200 killed. In each village the Little Swords demanded food, pork, chickens, and guns. What had started as a feud ended up as banditry in the name of self-defense. Buck appears to argue that an innovative approach to solving the banditry problem and in-feuding between villagers is to establish formal, government-recognized self-protection societies associated with a credit society (as had been done with the Jiangsu Government Rural Bank) and use the credit line to purchase weapons for self-defense. Ultimately both the Big Swords and the Little Swords developed individual credit societies; although for several years, while the feud played out in combat, the formal lenders could not collect upon the loans. When the feud finally ended both groups apparently paid their debts in full.

[26] Buck, J.L. October 13th, 1928. The Big Swords and Little Swords Clash. *China Weekly Review* 46: 13–214.

Calamities also led to banditry. In Gansu, the famine provided conditions for the rise of a formidable "Mohammedan" bandit force of some 35,000 soldiers led by 18-year-old Major Ma Chung-Yin. Ma intended to eradicate the provincial warlord force of Marshal Feng Yu-hsiang, the leader of the Kuominchun, who was having problems of his own with General Chiang Kai-shek. In reality, the Mohammedan bandit force leaned more toward plunder and burning than administration.[27] Elsewhere in Gansu another Muslim uprising was reported, with a force of 20,000 associated with Sichuan warlord Wu-pei-fu (and other warlords) taking control of all of southern Gansu except Tsinchow and Kongchang.[28] In a horrific squabble between Chinese and Muslims in Gansu nearly 10,000 Muslims were killed and many hundreds of Chinese. In May 1929, rebellious Muslims under a warlord attacked Chinese and Tibetans, killing 80 Chinese in the Old City, followed by another 700 killed and 100 drowned above Lupasi. In retaliation Chinese entered the Old City, killing all Muslims unable to flee. When the Muslims returned in August they were separated, with all men between the ages of 15 and 50—some 2996—led outside the city gate and executed.[29]

Elsewhere the seemingly independent roving armies reached Hunan where General Lu Tih-ping was forced into exile at Nanjing. Lu's provincial forces were spread throughout Hunan on bandit suppression when, on February 28, a large military force dispatched by Wuhan Central Political Council entered the province on the grounds that General Lu was tolerating bandits, sympathetic to communists, and abused financial power. Lu could not mobilize a defense and did not want to plunge the war-torn city of Changsha into further turmoil.[30] By March 1929, the situation in Shandong was deteriorating for farmers, particularly in the eastern part and around Ankiu and Tsingchow. The Japanese protection zone along the Kiaochow–Tsinan (Jinan) Railway had formed a barrier that the nationalist troops could not cross from the west and which warlord General Chang Chung-Chang's bandit soldiers roamed with impunity

[27] Howard, H.P. June 1st, 1929. Famine and "Mohammedan" Banditry Again Devastating Kansu. *China Weekly Review* 48: 17.

[28] Editorial. March 2nd, 1929. Moslems Start another Rebellion in Kansu. *China Weekly Review* 48: 10.

[29] Editorial, October 5th, 1929. Moslems Massacred in Kansu. *China Weekly Review* 50: 199.

[30] Editorial. March 2nd, 1929. Hunan General Flees Under Pressure of Wu-Han Council. *China Weekly Review* 48: 10.

to the east with no interference from the Japanese. The bandit soldiers had split apart from the main force and set up small feudal locales over claimed territories in which many atrocities occurred. Farmers were robbed and when there was nothing to rob they were executed. Rape was rampant, with bandit soldiers billeted in farmhouses. Many women committed suicide rather than be raped. Desperate farmers rose up against the bandits but were no match for them and were killed in the hundreds and thousands, with entire villages leveled.[31]

In October 1929, starting around the 8th, 1600 villages were destroyed by fire, 20,000 people killed, and 10,000 carried away for ransom by bandits. The bandit force, 10,000 strong, attacked with impunity since the villagers had obeyed a central command to disband local militia. The bandits were eventually defeated by forces from Nanjing and Kaifeng.[32]

The year 1929 also saw a rise in communist activities, although the communists were treated more as a bandit force than a political threat. But by August 1929 the communists had been very active in the southern provinces. In some instances the communists kidnapped people for ransom. For example on September 30, 1929, 13 missionaries were captured from the Dominican Mission by Zhu De and Mao Zedong, eventually being released after a $10,000 ransom was paid.

Communist forces were reportedly active in western Fujian, northwest Guangdong, and parts of Jiangxi. Liencheng in western Fujian was taken by Zhu De and Mao Zedong on August 6, 1929; the Suishui district of Jiangxi was taken on August 4 with considerable loss of life and property. On September 23, 1929, the communists under Zhu De recaptured Shanghang. The communist forces did not appear to have any stable hold, for as one place was taken another had to be abandoned. However, in the abandoned areas the old land titles and contracts between landlords and farmers were destroyed. As the (liberal elements of) the nationalist Kuomintang (KMT) pushed communist forces out, they did not necessarily restore the old system of land tenure but actually reconstructed the districts on a new and equitable basis, preventing the landlords from "regaining their old position of privilege and exploitation."[33] By March

[31] Chung, S.K. March 30th, 1929. Ten Million People in Shantung Doomed to Death. *China Weekly Review* 48: 212.

[32] Editorial. October 12nd, 1929. Bandit Raids in Honan. *China Weekly Review* 50: 235.

[33] Editorial. August 17th, 1929. Chinese Communist Armies Still Active in the South. *China Weekly Review* 50: 530; Editorial. September 28th, 1929. Communists Recapture

1930 and through 1933 communist activity was reported in Henan, Hubei, Jiangxi, Fujian, and Guangdong. It was not until March 1930 that editorials finally concluded that the communist forces were likely a permanent force. In many instances the communists would establish "soviets." They would stamp out all vices when they took over a town, including gambling and opium and sometimes even tobacco, but they were not very tolerant of any bourgeoisie resistance. In April and March 1930, some 5000 uncooperative landowners were executed. Other "oppressors of the people" were either exiled or shot. Religions and idols were banned; however, reports from missionaries and other foreigners testified that looting was not rampant.[34]

The communist "Red Army" was made up of an amalgam of nationalist soldier deserters, warlords and bandit gangs defeated by the Red Army, and farmers. Recruiting farmers to the Red Army was an easy task. Corrupt magistrates and other officials as well as merciless militarists had stripped the farmers of their animals, deprived them of what little stores of grains they held, and left them without any worries at all except the all-absorbing grim alternative of life or death. In contrast, the communists protected farmers while they were growing their crops, eradicated local government, destroyed deeds, and confiscated hordes of food and redistributed it all, asking only for a small share in return. "Could there be a more appealing argument to a farmer who has been forced to sell his wheel barrow to pay his tax?" From these grim realities facing the Chinese farmer any change could only have been viewed as a gain, "*for simple arithmetic proves that no matter how much you subtract from zero, you get only zero. Having nothing left, they stood to lose nothing. Therein lay the strength of the communist appeal.*"[35]

The Red Army also worked alongside the so-called "Farmers Union," which was very aggressive at taking control of governments where Nanjing's influence was weak. At times the Farmers Union would lead combat raids, going ahead of the Red Army and carrying hoes and clubs. The majority of union members were from the landless or labor class and had nothing to lose under the structure of tenancy and sharecropping.

Shanghang. *China Weekly Review* 50: 161; Editorial. October 5th, 1929. Communists Capture Missionaries. *China Weekly Review* 50: 205.

[34] Editorial. March 8th, 1930. Communist Menace Gaining in the South. *China Weekly Review* 52: 75.

[35] Hunter, E. January 31st, 1931. The Seriousness and Extent of Red Armies. *China Weekly Review* 55: 322–325.

Landowners, small and large, were able to make a minimal living from renting out the land, but sharecroppers had to pay huge dividends in crop share to the landowners.[36]

In addition to bandits and warlords and communist insurgencies, China in 1929 and 1930 was faced with the prospect of two civil wars against the Nanjing KMT government under Chiang Kai-shek. These were largely political conflicts in which the warring parties were largely supporting the KMT but under the principles of KMT founder Sun Yat-sen rather than the republicanism of Chiang Kai-shek. In the southern provinces two forces arose and combined to challenge the government in Nanjing. These were largely from the Wuhan faction in Hubei province, which had an alternative government to Beijing. The second civil war was instigated by the Shaanxi Christian General Marshal Feng Yu-hsiang, who would ultimately pressure for yet another government in Peiping (Peking). Feng, once known as the Christian General, controlled a force of nearly 150,000, with the main force settled in Shaanxi. Feng was an officer during Qing (Manchu) dynasty (pre-revolutionary 1911) but secretly joined Sun Yat-sen, and moved his forces against the Manchus. Between then and 1929/1930 his forces were made up of republican loyalists with many excursions in battle and bandit control. It was argued that his men were probably the best of all military forces in China, and a force to be reckoned with. Feng was solidly in support of Sun's three principles, but after Sun's death and the rise of the Northern Expedition forces who championed a unified China he saw the revolutionary powers as being more interested in control than these principles. Nevertheless, outright civil war in Shaanxi, Shanxi, and Henan was averted (at least in 1929), but the southern rebellion was more militaristic.

As mentioned, the governing and military authority of Chiang Kai-shek had always been tentative, with armies comprised of seemingly loose alliances and even looser loyalty to the nationalist movement. In Hunan, three generals ousted an appointee of Chiang Kai-shek, which upset the general balance of power. Chiang Kai-shek, with 150,000 troops from the First Route Army in the area, started to send troops to Hunan. General Li Chung-jen, commander of the Fourth Route Army and part of the Wuhan or Jiangxi faction, also had 120,000 troops available. How troops behaved

[36]Godwin, F. June 7th, 1930. Red Banditry in Hupeh; Disorganization without Communism *China Weekly Review* 53: 24–25.

depended upon whether the various generals were allies or opponents within the KMT. The various generals still sought political and economic control over particular provinces. General Li Chung-jen was attempting to bring Jiangxi under his control while Chiang Kai-shek was still trying to extend his control over Hunan. It was clear that Nanjing was losing patience with the Wuhan faction and that if the Wuhan generals did not submit to Nanjing and the mandates of the central government another extended civil war would be inevitable. Indeed, a mandate against the Guangxi faction was issued later, on March 26, with a manifesto on why civil military action was required issued by Chiang Kai-shek on March 27, stating that what was at play was more than a simple fight between Hunan and Hebei but a challenge to the revolution itself.[37]

By April 1929, the situation along the Yangtze was very fluid, with a showdown between the Nanjing government forces under Chiang Kai-shek and the so-called Guangxi (Kwangsi) forces who were loyal to the nationalist government but not loyal to Chiang as leader. The Guangxi and Wuhan factions were being advanced upon in three directions by Chiang's forces from Nanjing but no real fighting had taken place, although Wuhan defenders reportedly placed mines in the Yangtze River downstream of Hankou. However, in a surprise move on April 4, a Guangxi supporter, General Hsia Wei, who headed the Seventh Route Army, switched sides to support Chiang and the First Route Army. Consequently, Guangxi troops refused to fight and turned toward Hunan to return to Guangxi. The conflict in Wuhan was deemed over, so much so that Chiang's orders to have Cantonese troops attack Guangxi troops was refused on the grounds that the retreating Guangxi troops were unlikely to engage in any further attacks. As Guangxi troops fled to Hankou, Chiang's troops moved in, causing many civilians to seek protection in the foreign concessions.[38] Ultimately the Guangxi rebels and so-called "Ironside" troops became a rebel force requiring a military response from Nanjing.[39] These were by no means minor skirmishes. In two separate battles in Hunan between the Guangxi or

[37] Editorial by JBP. March 23rd, 1929. Is it to be war or Peace on the Yangtze? *China Weekly Review* 48: 154.

[38] Editorial. April 6th, 1929. Hankow in Hands of Gen. Chiang Kai-shek. *China Weekly Review* 48: 226.

[39] Editorial. February 15th, 1930. Government Troops Rounding Up Rebel Bands in South China. *China Weekly Review* 51: 410.

"Ironside" forces and nationalist troops, there were 21,000 casualties of which 5000 were government troops.[40]

Militarism in China was not confined to banditry and civil war. For years prior to 1931 the Japanese had sought ways to increase their influence in Manchuria and China proper. On September 18, 1931, a Japanese force of about 40,000 occupied the Chinese territories in south Manchuria (north Manchuria at the time being controlled by the Soviets).[41] The occupied areas included southern Fengtien province, including its capital Mukden, which was also the seat of the Chinese political administration in Manchuria, and a considerable portion of Kirin (Jilin) province. Also occupied were key port cities of Newchang, Antung, and Changchun. Changchun was also the junction of the Japanese South Manchuria Railway and the Soviet China Eastern Railway. Some 30 million Chinese, a third of all Chinese in Manchuria, were now living in areas under Japanese occupation.

Within this territory lay Chinese railways that were adversely competing against Japanese and Soviet railways and also the Peiping–Mukden line which made up the main artery between Mukden (Shenyang) and Peiping (Beijing), south of the Great Wall to southern Chinese markets. In essence Japan had control over all ports and all rail and thus controlled the shipments of all goods into and out of Manchuria. The Japanese also took control of all telegraph lines and radio stations, so that all communication into and out of Manchuria went through them. In fact, the only news about the occupation came from Japanese sources and even foreign consular offices had to send communiques via Japanese lines and then have these retransmitted. Business interests in Shanghai and other business centers south of the Great Wall had no means of communication with interests in Manchuria.

What instigated the occupation is an interesting story. The Japanese accused the Chinese of blowing up a bridge on the South Manchuria Railway at Liutaiokuo station, north of Mukden. What ensued was a clash between Japanese railway guards and Chinese troops. This was

[40] Editorial. June 14th, 1930. Situation Deadlocked in Honan-Attention now Centered on Hunan and Shantung Fighting. *China Weekly Review* 53: 49–52; Editorial. July 5th, 1930. Yen His-Shan's Policies and Kuomintang. *China Weekly Review* 53: 169–172; Editorial. July 5th, 1930. Complete Collapse of Kwangsi and Ironsides Reported. *China Weekly Review* 53: 193.

[41] Editorial. September 26th, 1931. Japanese Military Occupation of South Manchuria. *China Weekly Review* 58: 127–132.

reported at 10:30 p.m. By 6:30 a.m. the walled city of Mukden was occupied and its Chinese forces and police disarmed. As daylight broke on the morning of Saturday, September 19, Japanese troops within Manchuria and from Korea were mobilized and on the move. In addition, Japanese citizens in Mukden had been pre-warned. Later it was discovered that the bridge had actually been blown up by Japanese militarists looking for an excuse for occupation; the connivance worked. By the 20th, riots had broken out in Kyutsekai and Lungchingson near the Korean borders, putting Japan in grave danger. By this action the militarists were able to convince Tokyo to expand military activities. The situation was already rocky, due to the murder of two Japanese militarists by Chinese or bandit forces. The two soldiers, Captain Nakamura and another, were traveling on civilian passports in plain clothes in Inner Mongolia and could have been declared spies on their discovery. Nonetheless, the so-called "Nakamura Affair" was settled amicably at the political level between Tokyo and China Foreign Office and as late as September 17 was not a cause for military action. Indeed, by September 19 reports were made that the murder of Captain Nakamura was by Chinese regular forces and that several Chinese officers were to be taken to Mukden for trial and execution. It was a report out of Tokyo that first suggested the bombing on the 19th was by a clique of young "hothead" Japanese militarists who were enraged by the murders and political solution, and sought military retaliation instead.

In actuality the Japanese often sought reasons to expand their military presence in Manchuria. Nakamura, traveling as a PhD researcher, probably was a spy and was executed according to the law at that time. The Chinese, seeing that the affair could lead to increased militarism and mobilization, evaluated the political calculus and concluded that the lives of a few officers were a small price to avoid an all-out military conflict. For those Japanese seeking this conflict a new excuse was necessary. Also the timing was impeccable. With the Nanjing government faced with military activism in Canton (Guangzhou) and by communists elsewhere, combined with the severity of the Yangtze and Yellow River Valley floods, there was not much China could do immediately except to protest. With regard to the Cantonese, Chiang Kai-shek urged a halt to actions to deal with the Japanese occupation, but with the communists Chiang Kai-shek would make no peaceful compromise.

The annexation of Manchuria was followed shortly thereafter by the Shanghai conflict.[42] The so-called "Shanghai Incident" of January 29, 1932, refers to the military clash between Japanese marines and China's Nineteenth Route Army. The incident itself was the culmination of a great many stressors and a general belief that the Japanese intended to occupy Shanghai as they had done with Manchuria.

The Manchurian occupation and the massacre of Chinese civilians in Korea led to widespread boycotts of Japanese goods and services, starting around July 1931. The boycotts were enforced by organizations created for the purpose, such as the Anti-Japanese Boycott Association, which enforced the boycott by picketing shops selling Japanese goods, harassing Chinese working for Japanese employers, and intimidating those Chinese buying any sort of Japanese goods or services. In some provinces and political jurisdictions legal authority was given to remove Japanese goods from stores and imprison sellers or buyers who violated the pact.

Protests by students denouncing Japan and calling for war with Japan over Manchuria led to increasing hostilities between Chinese and Japanese. Derogatory remarks were also made about the Japanese emperor, who was believed to be a son of Heaven. On January 18, 1932, five Japanese, including several monks, were attacked by a Chinese mob outside the San Yeh towel factory in the Shanghai district of Chapei. Two of the Japanese were wounded, with one of them, a monk, dying shortly thereafter. On January 20 a mob of 50 Japanese from the Japanese Youth Protection Society retaliated. Armed with knives, daggers, and clubs, they burned the towel factory and clashed with Chinese municipal police, wounding two and killing one. Three Japanese were shot by police, one fatally.

That same day Japanese residents sent a message to Tokyo requesting the Japanese government send warships and troops to completely suppress this anti-Japanese movement. These same Japanese clashed with police of the international settlement, wounding a British soldier. In the afternoon of the 10th the Japanese consul-general presented an ultimatum to the mayor of Shanghai concerning the events of the 19th including: (1) a formal apology

[42] Editorial. February 13rd, 1932. League Committee's Report on Shanghai. *China Weekly Review* 60: 335–336; Editorial. January 23rd, 1932. The Ban on Japanese Goods in Honan. *China Weekly Review* 60: 258; Editorial. March 12nd, 1932. The Chinese-Japanese Battle-Front at Shanghai and Geneva. *China Weekly Review* 60: 41, 42–63; Editorial. March 12nd, 1932. In the Wake of Chinese Conquest. *China Weekly Review* 60: 45; Editorial. May 7th, 1932. Proposed Sino-Japanese Shanghai Agreement is Innocuous Document. *China Weekly Review* 60: 317–318.

by the mayor, (2) the immediate arrest of assailants, (3) payment of solatium and hospital bills, (4) adequate control of anti-Japanese movements, and (5) the immediate dissolution of all anti-Japanese organizations involved in fostering hostile feelings, riots, and agitation.

On January 21st the mayor notified the Japanese consul that he could acquiesce to the first three points but the latter two could prove difficult. Later on the 21st the admiral of the Japanese fleet announced that if the mayor could not provide a satisfactory response then he, the admiral, was determined to take appropriate steps to protect the rights and interests of the Empire of Japan;[43] the text of a telegram from Tokyo to the Japanese consul-general implied that the admiral's ultimatum was explicitly referring to using the Japanese navy in order to dissolve the Anti-Japanese Society and give protection to Japanese residents. The mayor for his part was working with the municipal council to avoid any possible conflict and succeeded in closing down the Anti-Japanese Boycott Association. Nonetheless, by January 24, Japanese reinforcements arrived off Shanghai while Chinese troop reinforcements moved into Chapei.

The Japanese admiralty and the consul-general's ultimatum expired at 6 p.m. on January 28; before expiry the mayor delivered a note to the consul-general acceding to the demands in their entirety and this was deemed satisfactory. While the ultimatum was playing out, the International Settlement Defense Committee had issued a state of emergency, but did not lift this after the agreement for fear that China would be unable to follow through on its promises or that the Chinese citizenry would riot at the humiliation of succumbing. The state of emergency established certain defense rights within the boundaries of the international settlement but nothing was said about areas outside, including parts of western Shanghai, which included Hongkew Park, where many Japanese lived, and the Woosung railway station. Although this area was outside the international settlement it was not unusual for Japanese marines to have security posts in the region. Around 11 p.m. on January 28 the admiral notified the defense council that he was going to move a detachment of marines into the Chapei region and that Chinese troops should remove themselves to the western side of the Woosung railway station.

Chinese troops refused to move as demanded and fighting broke out between Chinese regular troops and Japanese marines. The Chinese held

[43] Editorial. January 30th, 1932. Once More, Japan, Beware!-Hands Off Shanghai! *China Weekly Review* 60: 267.

their ground at the station and were able to use an armored train against the Japanese. The Japanese in turn bombed the station and the train and then proceeded to bombard the surrounding buildings with incendiary bombs to remove snipers.

By early March 1932 the Japanese had not only advanced through Shanghai but had extended their reach to about 45 km from Shanghai, deep into the farming countryside. By then the battle had diminished into a number of skirmishes between the two forces; while the Chinese demanded that all Japanese troops remove to their warships, the League of Nations attempted to address the issues from Geneva. Surreally, the battlefield outside of Shanghai and the Chapei district is portrayed by a clutter of images of farmers tilling the fields after the vast armies had maneuvered over the area, and villages in which occupants worked at the end of a Japanese bayonet doing odd jobs while Japanese civilians burned outlying buildings not destroyed by their bombs. Farmers were killed en masse by both bombs and bayonets in this area. By May 7 1932, a Sino-Japanese truce was signed, ending hostilities between the Japanese and Chinese, with Chinese troops remaining in position and Japanese troops returning to the international settlement in Shanghai.

Despite Japanese aggression, internal conflict was not arrested in the remainder of 1932 and into 1933. Communist suppression in the south and Japanese aggression in the north laid waste to many farms and disrupted the agricultural economy in many ways, as has been discussed. The main focus was on communist suppression in the southern provinces, although civil wars in Shandong and Sichuan were also reported. Japanese hostilities in Manchuria extended to Jehol and the Great Wall but by May 31, 1933, an armistice had been signed, restricting the Japanese to the north of the Great Wall.

3.5 Geopolitics and Agricultural Conditions

What is largely unobserved in Buck's data are the provincial or regional geopolitics. Lease arrangements, farm tenancy, taxation, and so on are all factors with credible impact on agricultural productivity. For the most part these are systemic and cannot easily be captured by factor-specific variables. Rather, these are captured regionally by provincial instruments and more broadly by agricultural regions. These impacts were discussed by journalist Harry Paxton Howard[44]:

[44] Howard H. P. April 20th, 1929. The Problem of China's Unification. *China Weekly Review* 48: 324–328.

> *The farmer is the backbone of the Chinese people ... And it is they who are the worst sufferers from war when it invades their districts ... [The] fact remains that a policy in the interest of the cultivators is the soundest and most intelligent policy that can be pursued... A class driven to desperation by rent, taxes and usury is a basis for an impoverished economy and perpetual discontent ... A farmer policy is not a simple thing to be worked out particularly in the widely varying conditions throughout China... If the question was put to them as to what the government could do for them, the first thought of most of them would probably be ... 'Ah, please get off our backs!' ... The primary need is freedom from pillage ... a policy which will free cultivation from taxation.... The most essential features would appear to be (1) cheap credit, best of all in the form of co-operative credit unions; ... (2) Cheaper goods, ... (3) better prices for products, which could best be gained by cooperative marketing; (4) improved communications ...; (5) better seeds, new crops, other technical improvements ... At the same time, it must be remembered that the ultimate cause of famine is the lack of any margin in normal years. Most of the victims of famine in Shandong today have been sucked dry and rendered destitute by merciless taxation year after year.*

In addition to the absence of economic infrastructure, much of China was separated by such large distances and varied geography that trade amongst regions was complicated if not prohibitively costly. For example, by one account it took 22 days to travel from Shanghai to Chunking in Sichuan, a distance of about 1000 miles: four days up the Yangtze River to Hankou, another four to Inchang, and four more to Chongqing for a total of 12 days, and then on to Chengdu overland by chair in ten days. The road from Chongqing to Chengdu was actually a stony path about three feet wide over most of the distance, laid out in stone slabs about five inches thick, one foot wide, and three feet long. Transport was by professional coolies, who would organize gangs to carry man and luggage. Hilly and mountainous terrain made railroad construction difficult and expensive, which added to the economic isolation of the region, and despite the fact that, in 1929, some headway was being made in the construction of a motorway, the roads built to grade, village by village, were unpaved and almost impassable in rainy weather.[45]

[45] Nyhus, P. August 17th, 1929. Some Agricultural and Economic Notes on a Trip to Szechaun Province. *China Weekly Review* 48: 516–518, 521.

Furthermore, provincial governors, mostly former warlords or militarists of one sort or another, acted with near impunity in terms of allocating funds transferred to the provinces from the central government. For example, General Feng Yu-hsiang, the warlord premier of Shaanxi, had, on November 28, 1928, made an appeal on behalf of Shaanxi, Henan, and Gansu for $100 million to cover labor relief and direct aid, including infrastructure, well sinking, and irrigation. Brigandage, the formation of bandit groups, was on the rise as a result of the famine. Nanjing authorized $1.25 million (only 1.25% of amount requested) in December 1929 to be split among Jiangsu, Guangdong, Zhejiang, Anhui, Hubei, Hunan, Hebei, Sichuan, Fujian, and three eastern provinces, but it was not clear whether these were general allocations or specific to reported famines. Ultimately Feng received nearly $800 million from Nanjing for famine relief but used the money to instead buy airplanes and armaments. Apparently Feng saw no point in sending money into the countryside and then removing money in taxes, even though he did remit taxes. On the other hand, most farmers were in no position to pay taxes in any case, and any funds collected were nowhere near enough to support famine relief.[46]

As previously discussed, the communists used the problem of excessive taxation to win over farmers for military recruitment through the establishment of soviets. But by 1933 progress in other areas was being made, largely in response to the floods and observed conditions in agriculture by the University of Nanjing.[47] A series of credit relief efforts were organized by the School of Agriculture at the university. Included in this relief was the establishment of agricultural banks and, in fact, a central agricultural bank was considered.

The Nanjing-managed relief restored nearly 1 million mou of land using refugee labor (presumably with wheat payment) but in order to get farmers back to the fields an amortized loan system was established. Under this scheme farmers could borrow seed grain directly or money to purchase seed grain, fertilizer, or implements for the planting of winter crops. These loans were to be administered by Ningshu Agricultural Relief

[46] Howard, H. P. June 1st, 1929. Famine and "Mohammedan" Banditry Again Devastating Kansu. *China Weekly Review* 48: 17.

[47] Editorial. February 4th, 1933. Some Practical Applications of Farm Relief. China Weekly Review 63: 407–408.

Association "*to afford a permanent agency for promoting the agricultural and economic welfare of the rural population of this (Ningshu) region.*" One of the first projects was to colonize some uncultivated land in the Chuyung Hsien along the Ching Hwei River, which had been badly damaged during the 1931 floods. About 3500 mou was set aside to support about 150 farm families. Colonists would be provided with a loan to cover the costs of seed and other inputs, food, and material for a shelter for the first year and this would then be repaid within a three- to five-year period.

In addition, short-term loans and mortgages were issued to rice farmers facing low rice prices in the spring of 1932. Farmers would have had to sell twice as much of their stores than normal, and this provided an opportunity to bridge the price gap with a loan. To guarantee the loan, the farmers were required to deposit their cheap rice as a mortgage in one of 101 depots scattered across the area. It was reported that some farmers were walking as much as 30 miles with rice on their shoulders to get the loans. Most of the 3014 families assisted by the commodity loans held less than 30 mou of land. Larger farmers who had resources were not eligible. At most a farmer could deposit 1500 catties (1 cattie = 500 g or about 1 lb) and could receive up to $30 in credit ($0.02/cattie or $2/tan with 1 tan = 100 cattie). Interest was set at 1% and loan terms were from four to six months. When the loan was paid the farmer would remove the rice from the depot. An average loan was about $15. Farmers would be issued with a deposit receipt which itemized the transaction and its terms and this was the only record of the transaction. The deposit receipt was transferable and could be sold to a third party and the rice released to them so long as the loan was repaid with interest. If the loan was not repaid then the rice would be auctioned off.

In addition, the University of Nanjing reorganized cooperatives in the district to encourage wheat production using wheat cultivar #26, which was tested by the university to increase yields by 10–15%. This #26 seed was also loaned to farmers on the condition that they would follow certain cultivation practices in line with the cooperatives. Dean Lin stated "*These cooperative societies are, in our opinion, most important, and we hope to make them a permanent feature of our work in the Ningshu district. Think of the hundreds of things that we could do for soil and crop improvement, animal breeding, forestry, and for general rural betterment through such societies. The solution of China's agricultural problem in my opinion lies in having such co-operative societies properly organized and properly directed.*"

3.6 Summary

This chapter has reviewed the history of China over the period 1929 to 1933. The main events were the 1929 famine, the 1931 floods, communist insurgency and warfare with the KMT, banditry, and civil war. This environment, we believe, is critical to an understanding of Buck's data and to truly appreciate the economic determinants of agricultural productivity during that time period. This chapter provides a context for interpreting the circumstances and results of the chapters that follow.

CHAPTER 4

The Discovery and Restoration of Buck's Original Data

Hao Hu, Mikio Suga, Weiwei Zheng, Minjie Yu, and Funing Zhong

4.1 Introduction

In Chap. 2 we provided a brief biography on the work that John Lossing Buck initiated while at the University of Nanjing. This work culminated in two classic books: *Chinese Farm Economy* and *Land Utilization in China*. These books were based on a series of small- and large-scale surveys across China, with the most intensive being that used in the three-volume *Land Utilization in China* study. Even by today's standards the scope and range

H. Hu (✉) • F. Zhong
College of Economics and Management, Nanjing Agricultural University, Nanjing, China
e-mail: huhao@njau.edu.cn

M. Suga
Faculty of Economics, Hosei University, Tokyo, Japan

W. Zheng
Jiangsu Academy of Agricultural Sciences, Nanjing, China

M. Yu
Business School, Huzhou University, Huzhou, China

H. Hu et al. (eds.), *Chinese Agriculture in the 1930s*,
https://doi.org/10.1007/978-3-030-12688-9_4

of analyses was breathtaking, and widely cited by scholars. The work is viewed as the most important historical representation of China's rural society and economy in the early twentieth century.

The chapters in this book are drawn from Buck's original data. But up until the year 2000, it was believed that the original data was lost to history. As it turns out, this was not correct, and in 2000 a large stock of paper spreadsheets was discovered in Nanjing. This chapter describes the rediscovery of the data, how the data was preserved, the digitization process and methods of validation, and complications which arose. Ultimately this chapter sets the stage for what is to follow. But it also contributes to a broader discussion on how historical data of importance should be preserved, and the painstaking task of transferring information from paper spreadsheets to modern-day electronic spreadsheets, the minutiae of care required in validating numbers, the frustrations with discovering that not all records were preserved, and the final digestion of the reality that the luxury we have in the twenty-first century of abundant data with which to build econometric models does not necessarily apply to recovered data, regardless of how carefully it has been reconstructed. On the other hand, the validation processes involved in data reconstruction revealed more about Buck's study than what was originally appreciated. For example, the task of measuring output in kg/hectare in the present day was not so easy in China's Republican era. Every data point that Buck constructed had to be converted from local measures of area (the local mou) and weights and measures. While in the modern era a mou is normally understood to be a sixth of an acre, in Buck's era a mou was an area of land that grew a fixed amount of grain by weight. Thus a mou measured in a high production area would have a smaller mou than one measured in a low production area. Buck provided the conversions in his statistical appendix, but it was not until the team at Nanjing Agricultural University (NJAU) started compiling the data that the granularity of precision was fully appreciated. This chapter describes that process and in doing so provides the context from which our data is drawn and the variables available to us.

4.2 Discovering Buck's Data

By 1937, the work on *Land Utilization in China* was complete. Several weeks before the Japanese occupation of Nanjing, Buck left Nanjing to work on trade issues with the US government, returning around 1940 to Chengdu where the university had relocated. Buck remained in Chengdu (and the university) until 1944 when he returned to the USA. What happened to Buck's data during this time period is unknown. This section provides the background to the discovery of the archived data.

However, even before his *Land Utilization in China* was published in 1937, Buck's means and methods were being criticized. In the 1930s, Hansheng Chen and Junrui Qian published articles in the Chinese periodical *Journal of Chinese Rural Area* criticizing Buck for being too favorable toward the landlords at the expense of the farmer, and failing to uncover unfair allocations of land and tenancy exploitation in China. Indeed, Buck's methods followed the survey techniques promoted by George F. Warren at Cornell University and the Western approach to accounting for unpaid family labor. When all opportunity costs, including the cost of capital, were accounted for, Buck found that many farmer tenants were in fact better off than their landlords. This conclusion was not helpful to the communists, whose success in the field was largely based on propaganda claiming the opposite. By the 1950s the Marxist view of Hansheng Chen that agricultural woes were a problem of allocation, rather than economics, had taken hold and Buck's view was publicly denounced. For example, Yingbi Duan, in conversations with Funing Zhong, admitted that he was forced to criticize Buck's work without ever having read it or really knowing what it was about.

For his part, Buck was relatively neutral on the political front. In neither *Chinese Farm Economy* nor *Land Utilization in China* does he take any sides on conflicting political ideologies. In *Chinese Farm Economy* he notes that farmers were suspicious about the surveys because they feared the "*information they gave would be used as the basis for increased taxation, or in one case for seizure of land by a new militarist*" (p. 4). Whether this new militarist was Mao is not stated, but at the time Buck was writing *Chinese Farm Economy*, Mao was active in South China as a militarist, seizing land from landlords and reallocating to farmers. The greater sin was probably from Chap. 5 on land ownership and tenancy. There, Buck writes "*The attention of the National Government in China in its rural work has been directed partly toward the farm tenant problem, and the question of fair rent payment by the tenant ...*" (p. 159). Buck followed with arguments that the "*landlord is entitled to a fair income from his capital investment in his land and buildings, and that the tenant is also entitled to a reasonable reward for labor*" (p. 160). The return on capital, Buck argued, should also include a premium for risk, with a risk return greater for landlord entering a share agreement versus one receiving cash rent. This higher return for share rents would no doubt require the landlord to take a greater share of crops than the tenant—anathema to Marxists who would read this as being pro-landlord and anti-tenant. On the tenant side Buck argued that the imputed wage rate for labor and management should reflect the value of labor and management in its next best alternative—presumably in the wage-labor market. Since the wage-labor market was pitifully low, it would be easy for the Marxist to read into Buck's economic principle that farmers ought not to receive a decent wage. Buck went on to suggest that because landlords seeking prestige might pay too much for land, when an appropriate risk-adjusted discount rate was applied, "*This rightly would give the landlord only a meagre return and would force him to share the loss with the farmer*" (p. 161).

By 1937, when *Land Utilization in China* was published, Buck had become neutral on the subject: "*In this study no attempt has been made to appraise in detail the so-called agrarian situation which may be thought in terms of the political, economic and social relationships between farmers and other classes of society … Some reformers assign most of the Chinese agricultural ills to a faulty agrarian situation comprising such problems as farm tenancy, injustice in the settlement of legal questions and disputes, usury, exorbitant profits of middlemen and the like*" (p. 1). While clearly trying to stay out of the political fray, Buck could not have known in 1936 and 1937 of the impending war with Japan, nor the march toward liberation and the rise of the communists in 1949, so even these words could easily be interpreted as being insufficiently favorable toward the landlord class.[1]

Buck's political restraint simmered throughout the 1950s, although it is clear that he continued to monitor events in China's agricultural economy. His silence was broken in the lead up to, and the aftermath of, the Great Leap Forward. In 1962 Buck published *Food and Agriculture in Communist China*, along with Owen L. Dawson and Yuan-Li Wu. Buck's contribution on food grain production concludes "*The public has been misled by the boasts of the Chinese Communists and by their food grain production statistics for 1949–1958 which began with too low a figure in 1949 and then increased every year until they reported the greatly exaggerated amount in 1958*" (p. 48; see also Dawson p. 77). Buck's inquiry was sparked by observing a reported level of output for 1958 of 375 million metric tons

[1] See Stross (1986) on this point. Stross goes further, suggesting that Buck should have known and observed the geopolitical dynamic of the rise of communism and the significance that his studies on land tenancy would have.

of grain produced—almost twice that reported for 1957—an agricultural impossibility based on Buck's previous studies and other national statistics. This number was revised to 250 million metric tons in 1959, but even that number appeared highly exaggerated. The growth in agricultural productivity reported from 1949 to 1958 would be used to justify technical improvements in production and the deployment of technologies, many of which were reported to have failed. Buck surmised from various sources, including *Land Utilization in China*, that the actual 1958 production was likely closer to 160 million metric tons. At this level of production, and grain mix, the caloric availability was in the neighborhood of 2017 calories per capita/day compared to a minimum requirement of about 2800 calories/capita per day used in the nutrition chapter. If so, the Chinese were receiving less, rather than more, calories on a daily basis than before 1949. It appears, that leading into the Great Leap Forward, the communists concluded that in order to justify moving resources away from agriculture, they had to provide a convincing argument that food availability exceeded food demand. The opposite, of course, was true. Even in 1961 and 1962 as Buck was preparing this monograph he noted that the "downtrend" in agricultural production after 1958 "*led to extreme hunger and malnutrition, and possibly some starvation*" (p. 51). We now know through revisionist calculations that many Chinese died of starvation during this period.

At any rate, Buck's intrusion into Communist China's agricultural statistics, and the suggestion that the data were manipulated to support the Great Leap Forward, would have irked the Chinese government. Perhaps even more biting was that Buck used his land utilization data as a reference at a time when this work was being publically discredited.

Why this is important relates, at least partly, to our understanding of why it took so long to discover Buck's data. There is some evidence that the importance of preserving the worksheets was recognized. Matsudo Yoshiro, for example, notes that during the Cultural Revolution of 1966–1976 archivists at Nanjing made a special effort to preserve at least the final copy of the tables compiled by Buck.[2] These data would have

[2] Yoshiro, Matsuda (2007) "Rural Farm Surveys and Agricultural Census in China; A Review" in Kuribayashi (2007) "Restoration of Farm Survey" op cit. pp. 1–3. See also Zhou, Yingheng and Qun Su (2007) "Nanjing Agricultural University and John Lossing Buck" pp. 4–9.

Kuribayashi, Sumio (2007) "Restoration of Farm Survey of Rural China in 1930s and Comparison with the Present Sampling Survey of Chinese Farms" Final Report, Tokyo International University.

been a target since, as mentioned, the Communist Party of China (CPC) had long held, even from the days of publication in 1937, that Buck's favorable defense of the landlord–tenant relationship went against the party line that the landlords were exploiters of the tenants. This position was ripe for attack by the Red Guard. In fact by 1967 Nanjing University was a hotbed of Maoist zeal, and one of the first to organize students and professors of the humanities to follow Mao's farm-study dictum. This ultimately led to the "June 2nd Incident," which empowered students and faculty to take a hard, and even violent, line toward radicalism and counterrevolutionary thought.[3] Mao had protested against Western thoughts and Western buildings, and many of the Nanjing University Red Guard, CPC cadres, and students wanted to prove their loyalty. It is doubtful that during this period Buck's writings would have seen the light of day since any teacher using them would have immediately been attacked as a counterrevolutionary, but clearly there were those moderates within the school who sought to preserve the manuscripts nonetheless, and likely at great personal risk had they been found out.

The materials remained hidden for another 33 years until they reemerged in 2000. The survey data were kept at the library of the Agricultural Economics Department at the University of Nanjing, located in the main campus, now the main campus of the Nanjing University, until 1952. At that time, the University of Nanjing and Central University were merged into Nanjing University, while many faculties from the two universities were also merged, along with some units from other universities, and others separated into independent colleges. The Nanjing Agricultural College was established in this way, with the Department of Forestry included from 1952 to 1957, when it was further separated, becoming the Nanjing Forestry College (now the Nanjing Forestry University).

From 1952 to 1957, the Nanjing Agricultural College was located in Dingjiaqiao in northern part of Nanjing City. When it was further divided into two colleges, agriculture and forestry, the new Nanjing Agricultural College was moved to the current location in Weigang, outside the Zhongshan Men, near the east gate of the Nanjing City Wall.

During this time period, the Group of Agricultural History was separated from the Department of Agricultural Economics to form the Section of Agricultural History in 1956 under the newly established Chinese Academy of Agricultural Sciences, with headquarters in Beijing. As the

[3] Guoqiang, D. (2010). The First Uprising of the Cultural Revolution at Nanjing University: Dynamics, Nature, and Interpretation. Journal of Cold War Studies, 12(3), 30–49.

staff of the section were basically the faculty members of the Nanjing Agricultural College, the section was stationed on the campus of NJAU, along with the Department of Agricultural Economics. It is thus believed that Buck's data were transferred to the Section of Agricultural History in 1956.

During the "Cultural Revolution," Nanjing Agricultural College was moved to Yangzhou, and merged with the Northern Jiangsu Agricultural College into Jiangsu Agricultural College. Both the Agricultural Economics Department and the Section of Agricultural History were moved to Yangzhou. In 1979, after the ending of the "Cultural Revolution," Nanjing Agricultural College was separated from the Jiangsu Agricultural College, moved back to the current campus and resumed its original name. Our best understanding is that Buck's data were brought to Yangzhou and back to Nanjing by the Section of Agricultural History. They were packed in unmarked brown paper parcels. Few people knew what they were and almost nobody knew their historical significance.

The Nanjing Agricultural College was renamed Nanjing Agricultural University and celebrated its 70th anniversary in 1984, taking the establishment of the Faculty of Agriculture of the University of Nanjing, not the founding of the university, as its origin. In preparation for this event, the history of the college was to be officially documented. It was suggested by Professor Xixian Wang, then a Professor of Agricultural Economics and Director of the Library of the Nanjing Agricultural College, that Buck's data should be returned to the Department of Agricultural Economics to facilitate drafting the history of the department. Since then, the data were kept by the library of this department, which was later renamed as the College of Economics and Management.

Still, it seemed no one knew that Buck's data were kept there, or what were in the packages. The few people that knew of their existence had no interest in the data and therefore did not examine it, nor deem it important to mention its existence to the empiricists then populating the department. In 1986, Professor Xingsui Cao, who received his PhD in agricultural history from the section and knew of the data, met with Professor Renduan Chen of Nihon University, and told him about Buck's data a few years later, in 1988 or 1989. In 1992, Professor Cao met with Professor Tajima Toshio of the Tokyo University, who was conducting a rural survey in Anhui province. Cao told him about Buck's work in Anhui in the early 1920s and the survey data collected in early 1930s then stored at NJAU. It is quite likely that Professor Matsuda Yoshiro of the Tokyo International

University (TIU) and President of the Japanese Statistics Society obtained the information about Buck's data from either of these two professors.

At the request of Yoshiro, Buck's household microdata was uncovered around 2000. Most of the original manuscripts had been well preserved, but some of them were lost, incomplete, or illegible. Recognizing the importance of the data it became quite urgent to sort out and preserve these remaining materials.[4]

In November 2002, the College of Economics and Management, NJAU, and TIU collaborated on developing Buck's data, using a Grants-in-Aid Program of the Japanese Society for the Promotion of Science (JSPS) with Professor Sumio Kuribayashi as the principal researcher.[5] From NJAU the project was pushed by Professors Hao Hu, Funing Zhong, Yingheng Zhou, and Qun Su. Over the years countless undergraduate, Master, and PhD students have been involved in the project, but in the early years of the NJAU–TIU collaboration there was also an exchange of students, with Takashi Osato being the first from Tokyo to go to Nanjing, and Liu Wei Wei the first from Nanjing to go to Tokyo.

4.3 Moving Forward

The reconstruction efforts made limited early achievements due to exceedingly huge amounts of data and limited resources. This international partnership maintained collaboration from 2003 to 2006, then began to digitize the original data after receiving funding from the Oriental Library (Tokyo Bunko) in 2007, successfully scanning all original materials (see Figs. 4.1 and 4.2) by the end of 2008. Nevertheless, the follow-up work had to be suspended because of insufficient funds and manpower until 2011, when the Major Social Science Program of NJAU provided financial support.

Figure 4.1 illustrates the procedures involved in the original handling of the manuscripts. The upper left photograph depicts the spreadsheets as found. Each package was carefully unwrapped and prepared by recording and collating each document. Ultimately, to preserve the fragile worksheets from possible, and further, deterioration, each worksheet was photographed. In a few cases the photographic images might not have

[4] Kuribayashi, Sumio (2007) Preface to "Restoration of Farm Survey of Rural China in 1930s and Comparison with the Present Sampling Survey of Chinese Farms". Page viii.

[5] The JSPS project #15402020 was titled "Restoration of Farm Survey of Rural China in 1930s and comparison with the Present Sampling Survey of Farms".

Fig. 4.1 Preparing original manuscripts before digitization process

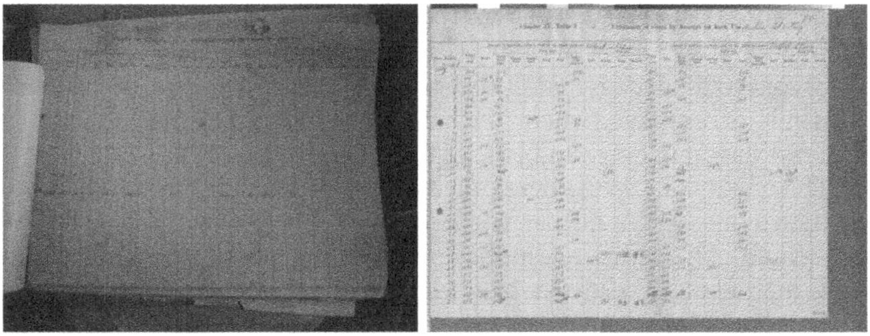

Fig. 4.2 An example of original manuscripts and scanned picture

been clear enough, particularly on faint images, and only in these instances were the original documents reexamined.

Buck mentioned in his books, that the first survey (1921–1925) covered 2866 households, 17 counties, and seven provinces, while the second one (1939–1933) included 16,786 households, 168 counties, and 22 provinces, covering the whole of China except the northeast, Tibet, Outer Mongolia, and Xinjiang. By 2008 there were about 400GB of scanned files, which contained all of the original materials preserved in NJAU, totaling 118 boxes and 24,956 pictures. All told, there were 189 table titles which were reduced to 86 tables after consolidating repeat titles. Ultimately, 172 counties were confirmed, four of which were part of the first surveys making up parts of *Chinese Farm Economy*, with the remaining 168 belonging to the second survey, which started in 1929.

Figures 4.3 and 4.4 provide two maps indicating (by black dots) all of the counties (*hsiens*) surveyed by Buck, and those surveyed by Buck and for which records on "Able Bodied Men" were recovered. There are two observations of note that become particularly important in the empirical works of this book. The first, relating to Fig. 4.3, is that Buck did not necessarily gather all variables in the survey at all locations. Why this is, is not clear, although Buck does note that methods of stratified sampling were employed. Thus in Fig. 4.4 the black dots record locations that were surveyed while the red points indicate locations for which there were records of labor. Likewise, in Fig. 4.5, for yield per mou, the red dots indicate locations for which data was available and recovered. For the most part, for any particular item the records were complete for those localities actually surveyed. However, the spartan nature of the data at the varied locations do not always overlap. From an analytical point of view this is important to understand because, as is evident in several chapters, the number of observations do not necessarily match the totality of data originally collected or recovered. This becomes evident in Chap. 11 on credit demand for example, where the addition of variables into regressions reduced the sample size and efficiency of the estimates.

The quality of existing data varied depending on table titles (i.e. investigation content). A further challenge was keeping track of each farm household. The actual questionnaires were not preserved. Rather, each table represented a particular item from the questionnaire, and these items were collated by farm, village, province, and region. Buck's stratification procedure held that in each village survey exactly 100 households were to be surveyed. These households had to be matched across all tables. Figure 4.6 illustrates the challenge. Each paper table represented a survey item at a particular locality. The farms were ordered so that farm 1 in Table XXX for *hsien* YYY was the same farm 1 in Table ZZZ for *hsien* YYY. For electronic digitization this ordering had to be preserved so that items across tables could retain order across multiplication and addition, not only for the same farms but also across farms. Figure 4.6 provides an

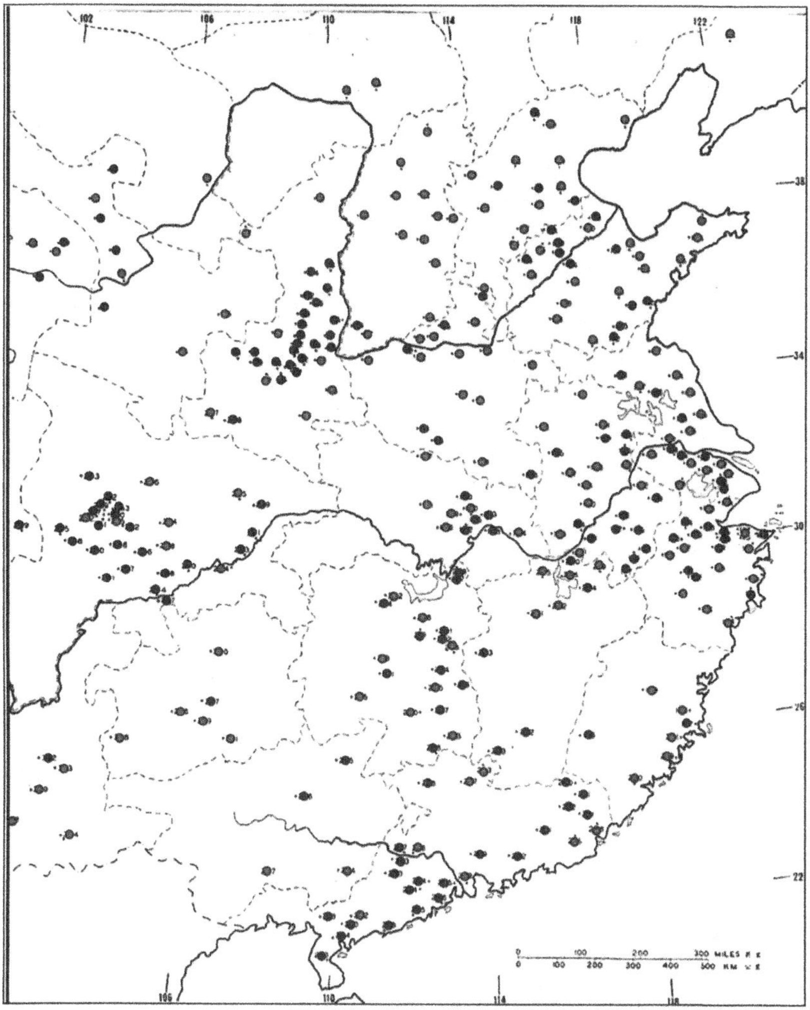

Fig. 4.3 Counties surveyed by Buck and counties with recovered data of any type

example of a paper spreadsheet ordered by farms for a particular *hsien*. Each original manuscript was not a questionnaire for a rural household, but detailed survey data, total, and grand total for a certain title (see Fig. 4.6), which could provide original information of each household too.

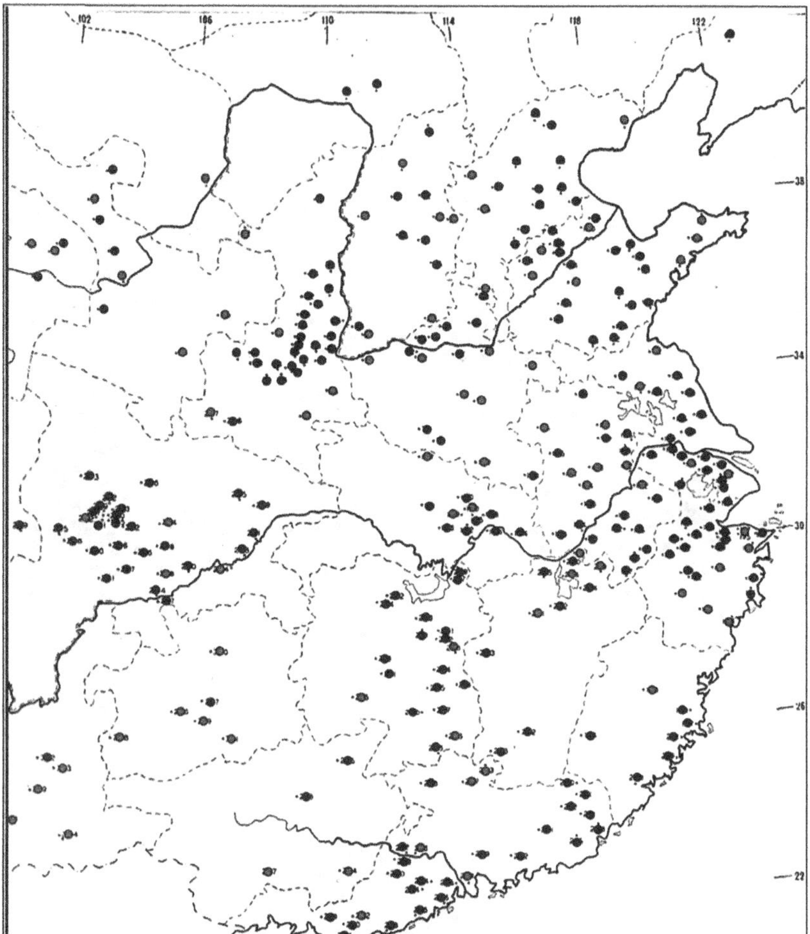

Fig. 4.4 Counties surveyed by Buck and counties with recovered data for chapter VII, Table 5 "able bodied men over 15 and under 60 years of age"

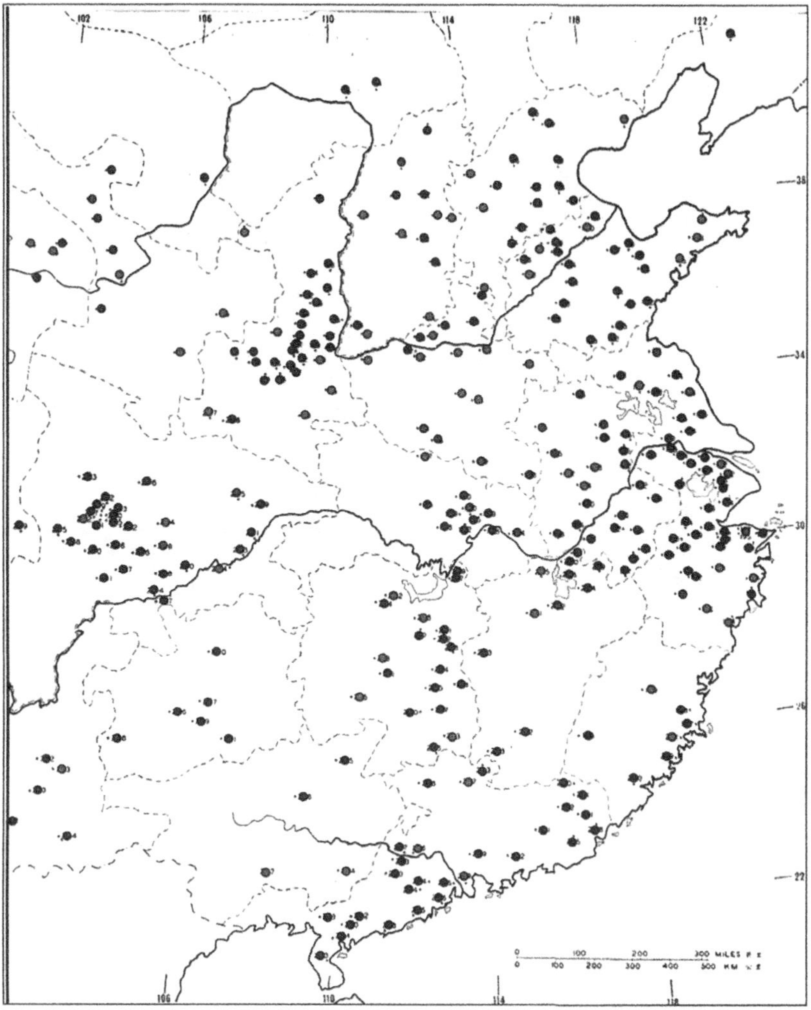

Fig. 4.5 Counties surveyed by Buck and counties with recovered data for chapter IV, Table 21 "Yield per mou (mu) of all crops (in ton and catties) (grouped by size of farms)"

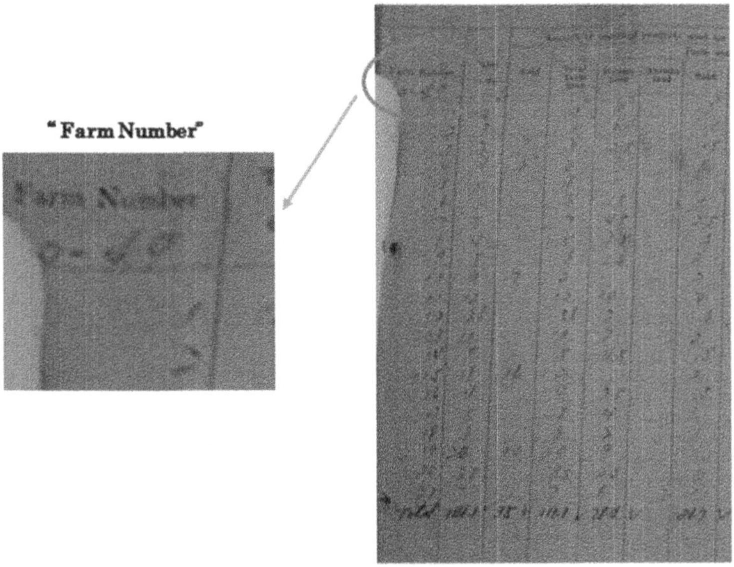

Fig. 4.6 Paper spreadsheet ordered by farms for a particular *hsien*

4.4 Main Work

4.4.1 Entering and Checking Data

The research team organized more than 80 undergraduates (second year or above) majoring in agricultural economics and over 20 postgraduates to enter data. The task required entering all information (including table title, region name, variables, data, etc.) to Excel in exact accordance with the scanned pictures. This is illustrated in Fig. 4.7. Each digital picture corresponds to an Excel sheet which was labeled according to the picture number. Correspondingly, each picture box was to match an Excel file as illustrated in Fig. 4.8. Meanwhile, the research team checked all entered data one by one at least twice, in order to avoid mistakes caused by carelessness. On this basis, we translated all English (including table title, region name, variables, etc.) into Chinese, to complete two sets of Excel files—an English version and a Chinese version.

On an initial scan of the manuscripts it appeared that 72 separate tables were included. However, how these fitted with the final books on *Land Utilization in China* was initially ambiguous and confusing. As it turns

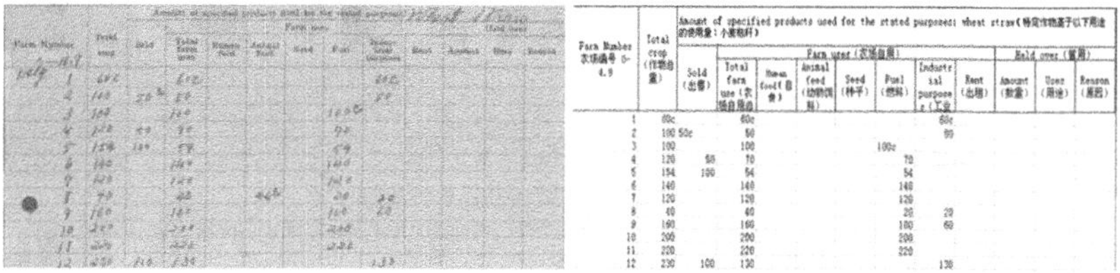

Fig. 4.7 An example of the digital picture and corresponding Excel sheet

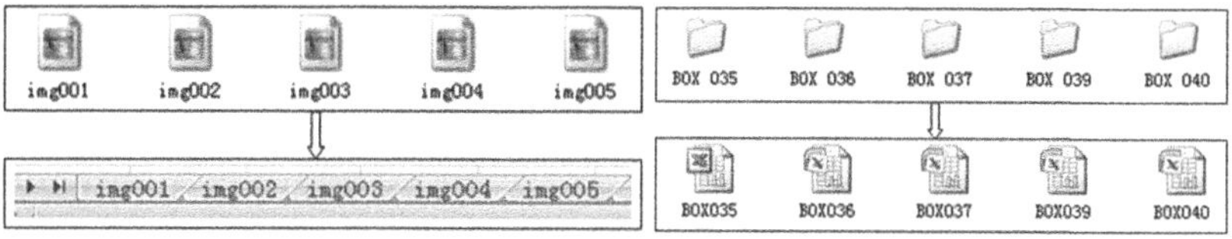

Fig. 4.8 Scanned files and corresponding Excel files

out, the titles and chapter identifications in the original documents were tentative, with the final chapters and tables being completed at the time of publication. For example, the original documents identified as "*Chapter IV Table 8a & 12*" was found to be "*Utilization of crops by amount for each use.*" In the published volumes "*Chapter IV*" deals with "*Climate*" and Table 8 is "*Causes and effects of famine within memory of informants.*" In the final volume there is a similar table, titled "*Utilization of crops by percentage for each use*," but this appeared as "*Chapter VII, Table 23.*" Ultimately, by examining tentative titles and data points each table type was matched and reworded to conform to the tables and chapters in the final published volumes.

As the digitization process continued, it was discovered that there were actually two table types: household tables and county tables. The former, 25 in total and reported in Table 4.1, included farm numbers and recorded information for each agricultural household. The county (*hsien*), totaling 61 (provided in the Appendix), showed only the county names and provided specific features for each county, including measures of local mou and other weights and measures that became indispensable in matching Buck's published data to those in the statistical volumes. County data were in line with those recorded in *Land Utilization in China: Statistics*, published in 1937, and could easily be checked. The household data, on the other

Table 4.1 Titles of household tables available from digitized Buck data

No.	*Title*
1	Size of family (farm grouped by size of farm)
2	Able-bodied men (over 15 and under 60 years of age)
3	Proportion of all farm and subsidiary works performed by family and hired labor, by men, women, and children
4	Amount and distribution of livestock
5	Relation of size of farm to crop mou per labor animal unit
6	Farm area devoted to different uses grouped by size of farm
7	Number and area of graves in farms
8	Number, distance, and size (crop area in local units) of plots and fields
9	Proportion of farm area rented
10	Crop mou area per farm (farm group by size of farm)
11	Number of mou of crop area devoted to various crops
12	Amount of fertility produced on the farm
13	Amount and kind of fertilizers applied per mou
14	Changes in the use of fertilizers
15	Changes in kinds of fertilizers used
16	Yields per mou of all crop (in ton and catties) (grouped by size of farm)
17	Most frequent yield per mou of byproduct of important crops (in catties) (groups by size of farms)
18	Most frequent yield of important crops by soil types and irrigations
19	Utilization of crops by amount for each use
20	Utilization of crops by percentage for each use (by products)
21	Utilization of crops by percentage for each use (main products)
22	Utilization of minor crops by amount for each use
23	Savings
24	Credit and indebtedness
25	Special expenditures (by size)

hand, were checked via Excel functions to match those in the statistical volume. Given available data and its completeness, database construction concentrated on 25 household tables as listed in Table 4.1.

It is also worth mentioning that all information, describing the same content for households from same county, were scattered over several Excel sheets rather than only one. Some, for example crop type, were distributed across more than 20 Excel sheets. Likewise, the same content for households from different counties was also dispersed across as many as 20 spreadsheets. Ultimately, for the purposes of research convenience and data manipulation, it was necessary to organize data reflecting similar content together.

4.5 Integrating Data

The integration of the data into a working form needed to satisfy two criteria. The first was that the data structure had to be maintained at the granular level so that the final digitized spreadsheets could be subtotaled and totaled to match the abacus calculations of Buck and his team. The second requirement was that the final data sheets had to be accessible and mergeable so that farm observations and multiple variables could be constructed as a single-panel data set. To achieve this, a five-step integration policy was put in place. The integrating steps are depicted in Fig. 4.9 "Five steps for integrating data," and summarized as follows:

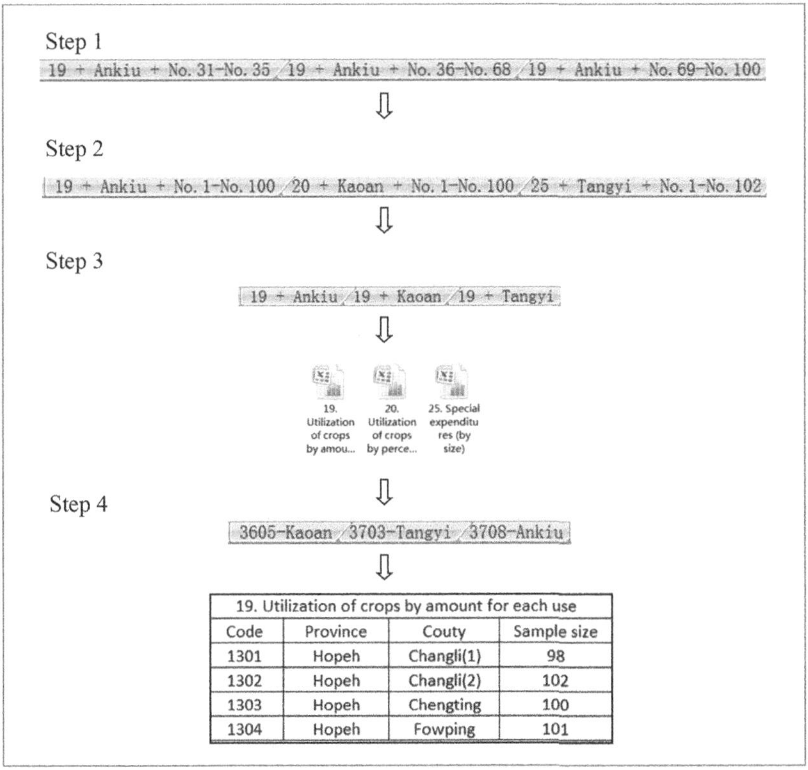

Fig. 4.9 Five steps for integrating data

- Step 1, rename Excel sheets following the format of "title number + county name + farm number" (e.g. 19 + Ankiu + No. 31–No. 35).
- Step 2, integrate different Excel sheets with same title number, same county name, but different farm numbers into one sheet, and label this according to the above format (e.g. integrate 19 + Ankiu + No. 31–No. 35, 19 + Ankiu + No. 36–No. 68, and 19 + Ankiu + No. 69–No. 100 into one Excel sheet, and name it 19 + Ankiu + No. 1–No. 100).
- Step 3, integrate different Excel sheets with same title number but different county name into one Excel file, and name it as "title number. Title name" (e.g. integrate 19 + Ankiu, 19 + Kaoan, 19 + Tangyi, and so on into one Excel file, and name it "19. Utilization of crops by amount for each use").
- Step 4, code uniformly for every county (see Appendix), then rearrange sheets in ascending order and make a catalog covering title number, title name, county name, and sample size for all Excel files.

4.6 Identifying and Repairing Problematic Data

There were three main types of problematic data: value identification, spelling, and dialect. For the most part these could be addressed in context, by returning to Buck's written work, and trial and error. In many instances several modern tricks could be used. For example, a good guess could be entered into the spreadsheet and the final total could be compared to Buck's tables, or Excel Goal Seek could be used to find ambiguous numbers to meet subtotals and/or totals recorded in the actual tables.

4.7 Values

These value issues manifested as indecipherable writing, text having being revised several times, or missing information. Here we illustrate some typical examples.

Example 1: The decimal less than 1 was expressed as "." in manuscripts, so sometimes it could not be determined whether this value was a decimal less than 1 because of blurred or faded "." (see Fig. 4.10). Sometimes, it was difficult to distinguish similar values in the manuscripts. For example, in Fig. 4.11 what appears as 11 is actually 17, and what appears as 2320 is actually 2520. At times these were difficult to catch. However, every table was designed to compute subtotals and totals and compared to Buck's

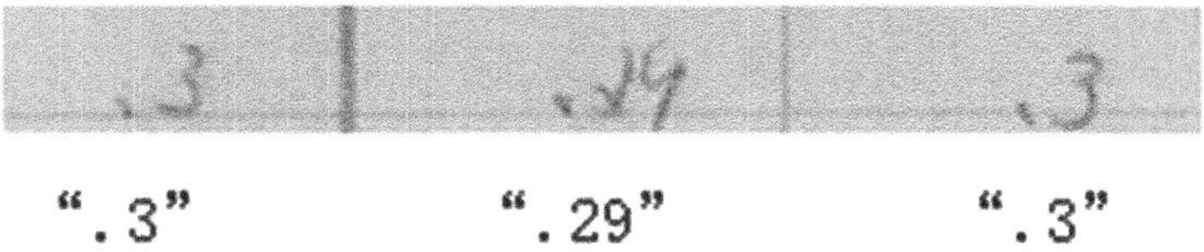

Fig. 4.10 An example of value with faded decimal

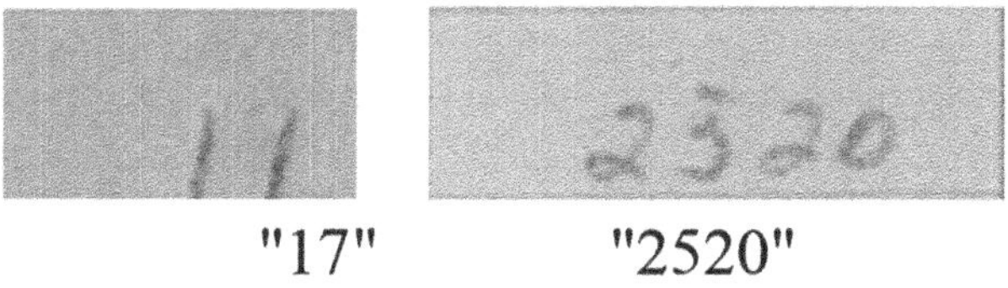

Fig. 4.11 An example of number ambiguity

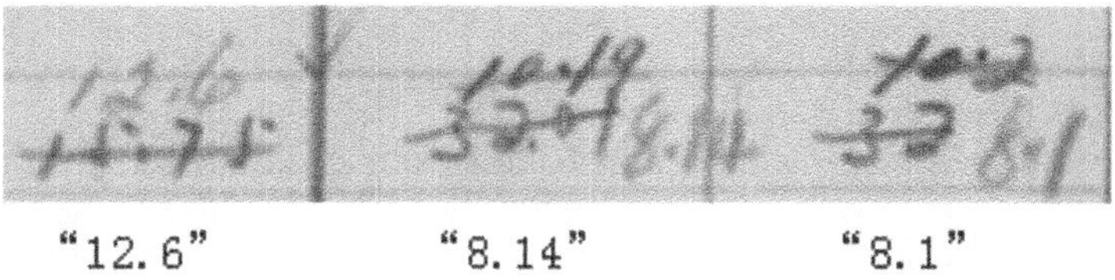

Fig. 4.12 An example of overwriting and corrected data entry

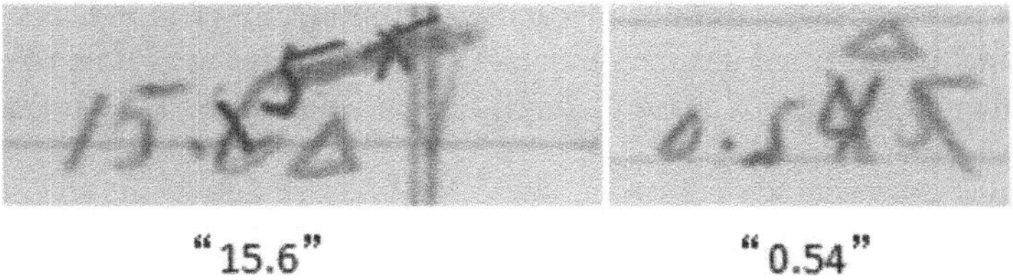

Fig. 4.13 An example of overwriting and corrected data entry

statistical volume. If a discrepancy was found, the team would return to the original paper document for reexamination. Figures 4.12 and 4.13 present a different problem, in which numbers were scratched out and rewritten. As mentioned, ambiguous numbers could be recovered by matching subtotals and totals in the paper spreadsheets and cross-validating with the statistical volume (see Fig. 4.14).

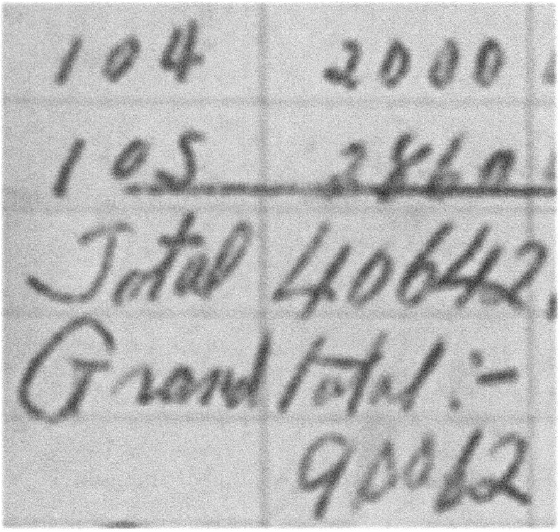

Fig. 4.14 An example of written totals and grand totals

4.8 Spelling

Due to different writing habits, the handwritten English letters with totally different appearances confused those data entry staff whose mother tongue was not English, such as "S" and "R," "a" and "o." These kinds of spelling problems usually appeared in the names of crops. For example, the record in Fig. 4.15 represented the name of a byproduct of an important crop, which was not easily recognized. Since this byproduct came from an important crop, we carefully checked every important crop in this county, and finally confirmed it as "rape stalk." By analogy, we were able to recognize illegible names of crops through comparison among relevant tables.

4.9 Chinese Dialects

Buck made a habit of ensuring that teams sent into local areas were familiar with local languages, customs, and dialects. However, on translation in traditional Chinese, these were at times indecipherable. For example, as shown in Fig. 4.16, the written Chinese characters actually represented the name of an important crop in Teian, Jiangsi. While the Chinese characters were identifiable, it was unclear as to what they represented. Returning to Buck's book we could infer the meaning in this dialect was turnip. But this method could only be applied to the names of important crops—the names of minor crops were still unsolved, such as "秋子" (*Qiuzi*), "土瓜" (*Tugua*), and needed to be investigated later.

Fig. 4.15 An example of ambiguous writing for "rape stalk"

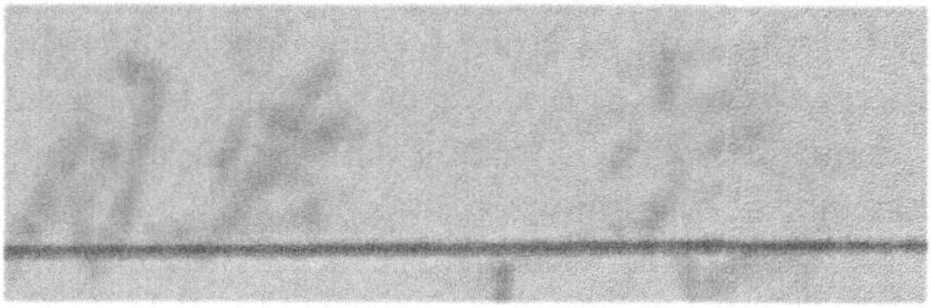

"(月虫)蓉"

Fig. 4.16 An example of ambiguous local dialect for "turnip"

4.10 Validating Household Data

The exactness required in validation to ensure that the digitized data matched in all measures to the summaries in Buck's statistical volume took considerable time, and considerable frustration. Once completed, the next challenge was to match the household data to that reported by Buck. Here, the research team explored whether exactly corresponding relations existed among different tables for the same number of households from the same county. First, we checked the sample size of each county to ensure they were consistent across tables. Second, we sampled the same variables from different tables and compared their values for accuracy. For instance, there was a variable named "most frequent yield" both in Table 16 (Yields per mou of all crop) and Table 18 (Most frequent yield of important crops by soil types and irrigations), and after checking carefully, we concluded that these values remained identical for the same number of households in each

Farm Number	Land Ownership					Crop Farm use
	Ower	Part-owner			Tenant	Rent
		Ower	Rented	Total		
1	11.5					
2		2.4	10			
3	13.5					200
4			15			
5	15.5					300

Farm Number	Land Ownership					Crop Farm use
	Ower	Part-owner			Tenant	Rent
		Ower	Rented	Total		
1	11.5					
2		2.4	10			200
3	13.5					
4			15			300
5	15.5					

Note: Left chart was wrong, while right chart had been modified.

Fig. 4.17 A comparison between land rented and rent amount

county. Finally, we compared relevant variables from different tables, such as production input from Table 13 (Amount and kind of fertilizers applied per mow) and output from Table 16 (Yields per mou of all crop), area rented from Table 9 (Proportion of farm area rented) and rent amount from Table 19 (Utilization of crops by amount for each use), and confirmed that there was a one-to-one correspondence between these relevant variables. By continually cross-checking data within and across tables, and under addition and subtraction where appropriate, we were able to validate the digitized data set as a near-complete replicate of Buck's data as reported in 1937. But this was not always straightforward. For example, as can be seen from the left-hand chart of Fig. 4.17, the information on area rented (in Table 9) did not match the rent (in Table 19) for household #5, so an error had obviously occurred in the manuscripts. We confirmed that the information of area rented was correct by comparing it with the total farm area from Table 6, and thus modified the data in Table 19. This is shown in the right-hand panel of Fig. 4.17.

In addition, the research team checked the mined data and the statistics in *Land Utilization in China: Statistics* to affirm whether both were perfectly consistent, and whether there were missing samples. We began with an inspection of table titles, and found that the table titles in mined data were included in the book, but not vice versa. Therefore, we figured out that missing materials did exist since the information relating to climate, nutrition, prices and taxation, marketing, the standard of living, and population could not be found in manuscripts. We then checked the sample size of each county, to confirm correspondence with the numbers reported in the statistical volume. Where appropriate we also looked for conformity in statistical measures such as averages, median, or percentages of a certain county calculated from mined data just equaled to the values in the book. For the data and variables available to us we found that the household samples were perfectly complete. Missing data on a number of sections in the statistical volume such as climate, nutrition, population, and so on could not be reconstructed.

4.11 Converting Local Units

The measurement system in early twentieth-century China was not yet unified. The units recorded in manuscripts were local ones, which were converted into metric units in *Land Utilization in China: Statistics*. To ensure the comparability between data, it was essential to reconcile the raw data with the appropriate conversion rates of measurement.

4.12 Unit of Area

The unit of area used in the manuscripts was "mou" (or mow). However, the measurement of mou as an area was based on the amount of area required for a given weight of production, and because production differed across the various regions of China, so too did the measure of mou. By contrast, the unit of area in Buck's book was "hectare." Fortunately, Buck listed the detailed information of how many local "mou" were equivalent to a "hectare" for each county (expressed as α), which was an excellent conversion guide for us.

4.13 Unit of Yield

The amount of yield was expressed by units of "T," "C," "O," "P," and so on, some of which were units of weight, such as "C" and "O," and some of which were units of volume, like "T" and "P." There were regional variations in these units of yield, which were unified as "kilogram" or "quintal" in the book. We only obtained the conversion rate of local "C" into "kilogram" for each county which was listed in Buck's book. But how could we deal with the other ones?

Here, we take "T" as an example. After careful investigation, we found that, in every county, there was an average yield per unit of area of each important crop in the manuscripts, indicated by Y, whose unit was T/mou, and there was a corresponding value in *Land Utilization in China: Statistics*, represented by X, whose unit was quintal/hectare. Therefore, the research team established a formula of conversion rate between "T" and "kilogram" (β) as follows:

$$\beta = \frac{X\left(\text{quintal / hectare}\right)\times 100\left(\text{kilogram / quintal}\right)}{\alpha\left(\text{mow / hectare}\right)\times Y\left(\text{T / mow}\right)} \quad \text{(Unit: kilogram/T)}$$

According to this formula, the other units of yield could be translated into kilograms as well.

In order to verify the correctness of this method, we randomly selected several values whose unit were C/mou, and calculated the conversion rate of “C” into “kilogram” based upon this formula, and found them to be consistent with those in Buck’s book. We then used this method to convert local units of yield as appropriate.

4.14 Unit of Distance

The variables of distance, such as “distance of the farthest plots,” “average distance,” were not clearly marked units, while the unit of these corresponding values in *Land Utilization in China: Statistics* was “kilometer.” Therefore, we could have the conversion rate between this “unmarked unit” and “kilometer,” and unified all distance values as kilometers.

4.15 Unit of Currency

The units of currency in manuscripts were “D” or “$,” which were “Yuan” or “silver Yuan” in Buck’s book. So, it was not a complicated process to estimate the relationship between “D” or “$” and “Yuan” or “silver Yuan.”

4.16 Summary

The data used in the various chapters of this book are based on the reconstructed and digitized data from John Lossing Buck’s 1929–1933 survey. In order to use these data effectively several goals were put in place. The first, of course, was to preserve the original paper spreadsheets or manuscripts. The difficult task was then transferring that data into spreadsheet form for use in economic and econometric research. This work has stretched across 15 years of labor, but with the processes in place we believe, and find no evidence to the contrary, that the digitized data are a clean representation of John Buck’s household survey.

The complete household database covered 17,203 households residing in 172 counties, and 22 provinces, including four counties (Kiangtu (1),[6] Ishing, Tungtai, Tienmen) surveyed from 1924 to 1925. The scope of data and variables are summarized as follows. We divided 25 tables into three different categories: major tables, minor tables, and general tables, depending on how crucial they were to agricultural production. These are

[6] In some counties more than one location were surveyed, they are numbered and expressed as (1), (2), and so on after counties.

Table 4.2 Tables in order of importance

Categories	*No.*	*Title*
Major tables	1	Size of family (farm grouped by size of farm)
	3	Proportion of all farm and subsidiary works performed by family and hired labor, by men, women, and children
	6	Farm area devoted to different uses grouped by size of farm
	9	Proportion of farm area rented
	13	Amount and kind of fertilizers applied per mou
	16	Yields per mou of all crop (in ton and catties)
	19	Utilization of crops by amount for each use
Minor tables	2	Able-bodied men (over 15 and under 60 years of age)
	4	Amount and distribution of livestock
	12	Amount of fertility produced on the farm
	23	Savings
	24	Credit and indebtedness
	25	Special expenditures (by size)
General tables	5	Relation of size of farm to crop mou per labor animal unit
	7	Number and area of graves in farms
	8	Number, distance, and size (crop area in local units) of plots and fields
	10	Crop mou area per farm (farm group by size of farm)
	11	Number of mou of crop area devoted to various crops
	14	Changes in the use of fertilizers
	15	Changes in kinds of fertilizers used
	17	Most frequent yield per mou of the byproduct of important crops (in catties) (groups by size of farms)
	18	Most frequent yield of important crops by soil types and irrigations
	20	Utilization of crops by percentage for each use (by products)
	21	Utilization of crops by percentage for each use (main products)
	22	Utilization of minor crops by amount for each use

provided in Table 4.2. The data completeness varied with different tables, for example, Table 1 contained 168 counties, Table 2 included 102 counties, and Table 13 only covered 87 counties. The total number of sample households which were contained in all major tables was 7853, located in 77 counties, 18 provinces. This number was reduced to 4881 from 48 counties, 14 provinces; and then 1432 households residing in 14 counties, seven provinces when the minor tables and general tables were added respectively.

Appendix: Counties and County Codes

Province	County	Code	Province	County	Code
Hebei	Changli (1)	1301	Jiangsu	Kiangning	3220
	Chengli (2)	1302		Kwangyun (Kwanyun)	3221
	Chengting	1303		TaiShie	3222
	Fowping	1304		Tungtai	3223
	Kiaocho	1306		Hwaiyin	3226
	Nankung	1310	Zhejiang	Tangki	3301
	Sushui (Hsushui)	1311		Tonglu (1)	3303
	Tsangshien	1312		Tonglu (2)	3304
	Tsing	1315		Tungyang	3305
	Ki	1316		Fenghwa	3306
	Tung (Tong)	1317		LingHai (Linhai)	3307
Shanxi	Anyi	1401		Lishui	3308
	Tsincheng	1402		Yuyao	3309
	Pingting (Pinting)	1403		Kahing (Kashing)	3312
	Showing (Sheoyang)	1405		Yungkia (Yungka)	3313
	Lin	1407		Shunan (1)	3315
	Ningwu	1409		Shunan (2)	3316
	Sin	1411		Tehtsing	3318
	Tatung (Tatong)	1412	Anhui	Fengyang	3401
	Tsinglo (Tsingloh)	1413		Fowyang	3402
	Wuxiang (Wusiang)	1414		Ho (1)	3404
	TaiKu	1415		Ho (2)	3405
	Tsingyuan	1417		Hefei	3406
Liaoning	Liaochung	2101		Tungcheng	3413
Jiangsu	Yenchen (1)	3201		Wuhu	3414
	Yenchen (2)	3202		Liuan	3416
	Yenchen (3)	3203		Su	3417
	Yenchen (4)	3204		Taihu	3418
	Kunshan	3205		Siuning	3419
	Wusih (1)	3206	Fujian	Putian	3501
	Wusih (2)	3207		Nanping	3502
	Wusih (3)	3208		Hweian	3503
	Wusih (4)	3209		Minhou	3504
	Wuchin (1)	3210		Lungki	3505
	Wuchin (2)	3211	Jiangxi	Tuchang	3601
	Wuchin(3)	3212		Nanchang	3602
	Kiangtu (1)	3214		Pengtseh	3603
	Kiangtu (2)	3215		Teian	3604
	Changshu	3216		Kaoan	3605
	Fowning (Funing)	3217		Fowliang	3606
	IShin (Ishing)	3219	Shandong	En (1)	3701

(*continued*)

(continued)

Province	*County*	*Code*	*Province*	*County*	*Code*
Shandong	En (2)	3702	Hunan	Changning	4316
	Tangyi	3703	Guangdong	Chaoan	4401
	Hweimin (Hweiming)	3704		Chungshan	4402
	Fushan	3705		Koyiu	4406
	Laiyang	3706		Kityang	4407
	Tismo	3707		Kukong	4409
	Ankiu (Anchiu)	3708		Moaming	4411
	Ishui	3711		Namyung	4412
	Ningyang	3717	Guangxi	Yungning	4503
	Sheokwang	3719		Jung	4504
	Taian (Tainan)	3720	Sichuan	Neikiang	5101
	Tsining	3722		Ta	5102
	Wei	3723		Suining	5106
	Yi	3727		Chongqing	5116
Henan	Loyang	4101		Fowling	5119
	Lingpao (Linpao)	4102		Hwayang	5120
	Nanyang (1)	4103		Mienyang	5122
	Nanyang (2)	4120	Kweichou	Anshun	5201
	Lingchang	4104		Pan	5202
	Taingyang	4105		Tingfan	5203
	Shangkiu	4106		Tuhshan	5204
	Kaifeng	4107		Tsunyi	5205
	Hiangcheng	4108	Yunnan	Yuki	5301
	Yencheng	4109		Yuangkiang	5302
	Cheng	4110		Tsuyung	5303
	Chi	4113		Yungjen (1)	5304
	Singyang	4115		Yungjen (2)	5308
	Tsiyuan	4119		Pinchwan	5305
Hubei	Tsaoyang	4201		Iliang	5306
	Kishui	4202		Mengtsz	5307
	Yincheng	4203	Shaanxi	Chowchih	6101
	Yummeng (Yunmeng)	4204		Sunyi	6102
	Chungtsiang	4208		Tingpien	6103
	Tenmen	4213		Weinan	6104
Hunan	Chen (1)	4303		Shang	6105
	Chen (2)	4304		Mien	6106
	Chengteh	4305		Chenan	6107
	Hengyang	4306		Yuling (Yulin)	6120
	Linsiang	4310	Gansu	Kaolan (1)	6201
	Sinhwa	4313		Kaolan (2)	6202
	Yiyang	4314		Pinliang	6203
	WuKang	4315		Tienshui	6204
Gansu	Wuwei	6208	Ninghsia	Ninghsia	6401
Tsinghai	Hwangyuan	6301	Suiyuan	Kweishui	6601
	Sining	6302		Paotow	6602

References

Kuribayashi, Sumio. 2007. *Restoration of Farm Survey of Rural China in 1930s and Comparison with the Present Sampling Survey of Chinese Farms.* Final Report, Tokyo International University.

Stross, R.E. 1986. *The Stubborn Earth: American Agriculturalists on Chinese Soil, 1898–1937*, 92. Berkeley: University of California Press.

Yoshiro, Matsuda. 2007. Rural Farm Surveys and Agricultural Census in China: A Review. In Sumio Kuribayashi (2007) *Restoration of Farm Survey of Rural China in 1930s and Comparison with the Present Sampling Survey of Chinese Farms.* Final Report, Tokyo International University.

Zhou, Yingheng, and Qun Su. 2007. Nanjing Agricultural University and John Lossing Buck. In Sumio Kuribayashi (2007) *Restoration of Farm Survey of Rural China in 1930s and Comparison with the Present Sampling Survey of Chinese Farms.* Final Report, Tokyo International University, pp. 4–9.

CHAPTER 5

Reliability of John Lossing Buck's Land Utilization Survey Data: A Preliminary Test of Grain Yields

Funing Zhong, Hao Hu, and Qun Su

5.1 Introduction

The publication of *Chinese Farm Economy* (1930) and *Land Utilization in China* (1937) by John Lossing Buck marked the starting point of modern research applied to China's rural economy. The significant contribution of the two books was twofold: (1) research methodology and survey procedures of modern social sciences were introduced and applied to the study of real world issues; and (2) large samples of rural households were covered in the surveys, particularly those provided in the second book. Compared to the current rural household survey of some 60,000 households conducted by the National Statistics Bureau in China, the coverage of around 40,000 households in Buck's survey was truly a remarkable research project and won worldwide attention, as it deserved.

To many scholars and policymakers, the large sample in conjunction with the theory and techniques of modern social sciences provides some

F. Zhong (✉) • H. Hu • Q. Su
College of Economics and Management, Nanjing Agricultural University, Nanjing, China
e-mail: fnzhong@njau.edu.cn

H. Hu et al. (eds.), *Chinese Agriculture in the 1930s*,
https://doi.org/10.1007/978-3-030-12688-9_5

comfort as to the overall reliability of the survey data. However, some academics have questioned the reliability of the survey in providing a true representation of general conditions. Indeed, as we reported in Chap. 3 on calamities and conflicts, it is likely that drought, floods, and conflicts could have had a significant effect on yields when measured relative to typical and best yields recorded. But that does not imply that the data do not truly represent yields and productivity, especially if in the longer term we view the various events described as being random in nature. Those concerns are our own, however. More generally, scholars doubtful of the representativeness of Buck's data have raised different arguments. For example, doubt arises from the fact that the survey was conducted by undergraduate and graduate students of the University of Nanjing in areas near their hometowns. The argument is that, as most of those students were from rich families in relatively rich areas, the sample was very likely to be biased. Therefore, the conclusions drawn from the research might be incorrect and the data not as valuable as many believe.

The suspicion has some grounds in logic, but is not necessarily true either. In this special case, 16,786 farms in 168 localities, and 38,256 farm families in 22 provinces, were interviewed during the four years of the survey (1929–1933). Most of the students might come from rich families, but they were not necessarily from rich areas, and the 100 families surveyed in each location were quite likely to cover different income groups. Therefore, the reliability of Buck's research, as well as the potential value of the data, require an empirical test and cannot be determined by logic alone.

Fortunately we have a set of statistical data of China's agricultural production for 1914–1949, compiled by the late Professor Daofu Xu (1983), which provides a preliminary means by which to test the reliability of Buck's data. It is quite reasonable to assume that rich farm families may have farms not only larger in size but also more fertile, so they may obtain not only higher total output but also a higher yield per unit of farmland. As Xu's book contains average yield data for many crops in each province between 1914 and 1949, we may choose and compare comparable statistics with those covered by Buck's survey. If the yields of Buck's survey are systematically higher than those provincial averages, the suspicion on Buck's data might be accepted, otherwise the reliability and value of the data would find some vindication.

5.2 Statistical Data of Chinese Agriculture, 1914–1949

Because Buck's survey was the first to deploy large-scale household surveys there are no equivalent surveys of such a comprehensive nature with which to make a comparison, let alone to validate the data surveyed. However, as mentioned, production statistics were collected for major crops in all provinces, which could be used to check Buck's household survey on one item: yield per unit of farmland. As rich farmers were likely to have more and better farmland, yields on their land were likely to be higher than provincial averages. On the contrary, if there were no significant difference between the provincial averages and Buck's household data, then the suspicion of upward bias is unlikely to hold.

The late Professor Daofu Xu of Nanjing Agricultural University (NJAU) (previously the University of Nanjing) compiled and consolidated all available data in his book *Zhongguo Jindai Nongye Shengchan ji Maoyi Tongji Ziliao* (Agricultural Production and Trade Statistics in Modern China) (1983). The acreage, total output, and yield per unit of acre were tabulated for grain, oil seed, cotton, and other major cash crops at provincial levels, along with data for major livestock production. The data came from different sources, with that for 1924–1929 from the Monthly Statistical Report compiled from reports submitted by county governments, while that since 1929 from the Agricultural Report sponsored first by the legislative body of the central government and then by the Ministry of Industry.

Of course, the official statistics at that time might not be accurate due to many reasons, including lack of skilled personnel and discrepancy in measurements. The latter might be a serious problem as data for different time periods came from various sources. Professor Xu did his best to consolidate data from various sources with appropriate approaches to convert all comparable data into the same units. There are still a lot of questions regarding the accountability of the official statistics during the Republican era; nevertheless, the yield data for the two time periods (1924–1929 and 1929 on) provide an independent data series with which we can test the reliability of data collected in Buck's survey.

5.3 Comparison of Yield Data

5.3.1 *Selecting Samples regarding Crops and Time Periods*

As mentioned, the original data of Buck's survey of farm economy covers 16,786 farms in 168 localities and 38,256 farm families in 22 provinces. The criteria for choosing crops and time periods are based on data availability. There are completed wheat production data for 1883 farm families in 18 counties in nine provinces, and rice production data for 1487 farm families in 14 counties in seven provinces. As other data are not as complete, only those that are used in our comparison. There are three types of yield data in Buck's household survey: the current yield, the most frequently reported yield, and the best obtainable yield. As the survey was conducted over four years, from 1929 to 1933, it is unclear as to whether Buck's yield measurements reflected the current yield in the calendar year or the most recent crop harvested, which might have been in the preceding year. To deal with this issue we use the three-year (1931–1933) averages from the official statistics in our comparison against the current yield provided in Buck. We also compare the ten-year (1924–1933) averages of the official statistics and the most frequently reported yields in Buck's survey to check if the comparison of long-run yields is consistent with that of short run.

5.3.2 *Consolidating Data for Comparison*

As described by Buck, all original data in his survey were in local units. To consolidate the data from various regions and for different crops, we made sure that all local units were appropriately adjusted so that a unified and standard unit of yield per mou to kg/mou matched that of the official statistics. After converting all figures into the same units, we then calculated the weighted averages of wheat and rice yields for each province in the subsample of Buck's survey, taking the acreage of wheat or rice in each farm as the weight. This is equivalent to calculating county averages using household acreage as the weight and then calculating the provincial averages using the county acreage as the weight.

5.3.3 *Results of the Comparison*

Results of the comparison are summarized in Tables 5.1 and 5.2.

Our results are interesting. What we find is that for neither wheat nor rice is there a systemic bias one way or another in Buck's data. In fact, in

Table 5.1 Comparison of wheat yields unit: Jin/mou

Province	*Current yield of wheat*			*Long-run yield of wheat*		
	Buck's	*Official*	*Difference*	*Buck's*	*Official*	*Difference*
Zhejiang (Zhejiang)	230.82	147.70	83.12	193.32	162.57	30.76
Hebei (Hubei)	98.14	187.00	−88.86	96.61	194.33	−97.72
Kiangxi (Jiangxi)	86.68	143.30	−56.62	87.07	145.77	−58.69
Jiangsu (Jiangsu)	123.58	185.70	−62.12	109.98	175.90	−65.92
Sichuan (Sichuan)	346.56	248.70	97.86	345.17	206.90	138.27
Anhwi (Anhui)	63.44	141.00	−77.56	66.60	154.33	−87.73
Henan (Henan)	103.72	146.00	−42.28	121.94	138.67	−16.72
Gansu (Gansu)	43.45	91.30	−47.85	38.30	155.10	−116.80
Shanxi (Shaanxi)	72.08	91.70	−19.62	83.24	120.57	−37.32

Note: Jin (sometimes referred to as cattie or catty) is a Chinese weight unit equal to 0.5 kg while mou is a Chinese area unit equal to 1/15 ha; current yield of Buck's survey is that recorded in the year of survey while that of official statistics is the three-year average (1931–1933); the long-run yield of Buck's survey is the recorded as most frequently obtained while that of official statistics is the ten-year average (1924–1933).

Table 5.2 Comparison of rice yields unit: Jin/mou

Province	*Current yield of rice*			*Long-run yield of rice*		
	Buck's	*Official*	*Difference*	*Buck's*	*Official*	*Difference*
Zhejiang (Zhejiang)	381.93	347.30	34.63	360.10	380.43	−20.33
Hebei (Hubei)	353.71	316.00	37.71	292.46	401.33	−108.88
Kiangxi (Jiangxi)	315.14	354.00	−38.86	290.28	370.00	−79.72
Jiangsu (Jiangsu)	236.06	365.30	−129.24	233.74	361.77	−128.02
Sichuan (Sichuan)	194.33	415.30	−220.97	197.91	413.77	−215.86
Anhwi (Anhui)	522.98	316.00	206.98	462.23	350.00	112.23
Henan (Henan)	138.10	246.70	−108.60	308.29	240.23	68.05

Note: Current yield of Buck's survey is that recorded in the year of survey while that of official statistics is the three-year average (1931–1933); the long-run yield of Buck's survey is the recorded as most frequently obtained while that of official statistics is the ten-year average (1924–1933).

most cases we find the opposite of what critics have suggested; that, if anything, Buck's data understates rather than overstates agricultural productivity. For both the current and long-run wheat yields, Buck's survey reported lower figures in seven provinces out of the nine-province subsample compared with the official statistics, with the directions of differences being the same for all the provinces. The comparison of rice yields shows the same trend with a small difference. While Buck's survey registered lower figures in four provinces out of the seven-province subsample for current yields, five reported lower figures for long-run yields in the same survey, and some provinces changed the signs of differences in the comparison. Other comparisons are rather sparse. Fei (1939, p. 201) notes that a typical rice household in Eastern China would yield six bushels of rice per mou (1 bushel = 551.12 jin) or about 330.72 jin/mou, which is consistent with Buck's findings. Myers (1970), taking measures in central Hebei, records wheat for 1930 of approximately 204.35 jin/mou, which appears to be higher than that reported in Table 5.1. Tawney (1932, p. 49), citing data for 1914–1918 for all of China, comes up with—after converting pounds to jin (500 grams) and acres to mou (1/6 acre)—97.98 jin/mou for wheat, 97.06 jin/mou for corn, and 264.60 jin/mou for rice. The wheat yield is at about the midpoint of those provided by Buck, but lower than the official numbers except for Gansu and Shaanxi. Likewise, the rice yield is more or less in the lower third of Buck's estimates and lower than the official numbers, except for Henan.

Though the samples are too small to undertake meaningful statistical analysis of precisely how yield and productivity measures were undertaken, the simple comparison conducted in this chapter suggests the opposite to the suspicion of upward bias in data collected in Buck's survey.

5.4 Summary

We would probably be incorrect in steadfastly holding that Buck's original data correctly measures crop yields in all regions and at all dates of the survey. However, with some aura of reserved skepticism, the assessment here suggests that we would not be incorrect in asserting that Buck's data is not inconsistent with other measures taken at various points throughout the Republican era. On the main results, the relatively lower yields in Buck's survey could be explained by the way of recording yields at the time. First, the yields were not the exact figures obtained by actual measuring. Instead, they were "reported" by farmers in rounded figures. Naturally, farmers might report their yields by rounding down, that is, by cutting the small fractions. As the interval of rounding is about 25–30 jin, the average yields might be underreported by 10–15 jin per unit of farmland. Second, the local units in measuring acreage were equivalent to 1/4 mou in many cases, which would raise the amount of underreporting to 40–60 jin/mou. If we further assume that farmers tend to underreport their production purposely, the actual yields on farms under the survey might be even higher compared with that calculated above.

This discussion provides a plausible explanation why the yields data from Buck's survey are below the official statistics. Another reason, as discussed in Chap. 3, is that the years 1929–1933 were tumultuous, with a number of calamities, catastrophes, and conflicts that did lead to diminished yields in some cases. Even so, with the simple analysis provided here we do not find support for the suspicion of upward bias in Buck's data.

Nonetheless, the various chapters in this book use direct measures from Buck's data. It is impossible to return in history and validate all measures, or observe how surveys were conducted at the local level. However, throughout the following chapters we attempt to be as honest as possible about the data used, its collection and measurement, adjustments made, and so on. Perhaps at the end of the day, the most critical testament to data accuracy is the logic of economic assessment and the robustness of results that make up the various story lines presented when the data is put to the test of modern economic thought and rigorous and various econometric approaches that were not available to Buck in the 1930s.

References

Buck, John Lossing. 1930. *Chinese Farm Economy*. Chicago: The University of Chicago Press.

———. 1937. *Land Utilization in China*. Nanjing: The University of Nanjing Press.

Fei, Hsiao-Tung. 1939. *Farmer Life in China: A Field Study of Country Life in the Yangtze Valley*. London: Routledge & Kegan Paul, Ltd.

Myers, Ramon H. 1970. *The Chinese Farmer Economy: Agricultural Development I Hopei and Shandong, 1890–1949*. Cambridge, MA: Harvard University Press.

Tawney, R.H. 1932. *Land and Labour in China*. London: George Allen and Unwin Ltd.

Xu, Daofu. 1983. *Zhongguo Jindai Nongye Shengchan ji Maoyi Tongji Ziliao [Agricultural Production and Trade Statistics in Modern China]*. Shanghai: Shanghai People's Press.

CHAPTER 6

Tenancy Issues in Northwest China During the Republican Era

Minjie Yu and Hao Hu

6.1 Introduction

Tenancy has been recognized as one of the most important relations of production in the traditional agriculture of feudal China. The system of tenancy has long been a research subject of academic circles. Taking account of tenancy disparities across regions caused by the existing regional heterogeneity of China, it is necessary to "explore historical data to restore the truth of history," and "especially strengthen research on different regions to discover regional characteristics and lay [a] foundation for general conclusions at a national level."[1] During the Republican era there were few studies on tenancy issues due to a lack of statistical data and documented records. Therefore, it is necessary to further explore the rel-

[1] Li Jinzhen (2011).

M. Yu
Business School, Huzhou University, Huzhou, China

H. Hu (✉)
College of Economics and Management, Nanjing Agricultural University, Nanjing, China
e-mail: huhao@njau.edu.cn

H. Hu et al. (eds.), *Chinese Agriculture in the 1930s*,
https://doi.org/10.1007/978-3-030-12688-9_6

evant characteristics of land tenancy on the basis of historical data. This chapter, using Buck's original household data, explores the characteristics of tenancy in the economic development of Northwest China from the perspectives of the tenant's behavior and the tenancy system. In doing so we can verify, or provide supplemental conclusions on, China's Republican-era tenancy issue. This is not only an exploration in economics and history, but also an important issue when comprehending the core of the traditional economy of modern China.

In modern geographic regionalization, Northwest China is defined in two ways: in the sense of administrative division and in the sense of natural division. The former covers three provinces and two autonomous regions (namely, Shaanxi Province, Gansu Province, Tsinghai Province, the Xinjiang Uygur Autonomous Region, and the Ningsia Hui Autonomous Region). This includes the vast regions to the west of Greater Khingan Mountains and to the north of Kunlun Mountain—Altun Mountains and Qilian Mountains. From the perspective of weather conditions, soil characteristics, crop varieties, and cultivation modes, which are closely related to crop production, homogeneity within region by natural division is likely to be quite relevant Buck (1930). In fact, Buck's regionalization for China's agriculture was consistent with the objectively different natural environments Buck (1937a, b, c).[2,3] Except for Xinjiang Autonomous Region, Buck's Spring Wheat Area naturally and geographically coincides with Northwest China as currently defined. Based on this geographic definition, and the data available from Buck's survey, we obtained sample data for a total of 610 farm households in Northwest China, including four localities in Gansu Province (Kaolan (1), Kaolan (2),[4] Wuwei, and Tianshui), one locality in Ningsia

[2] John Lossing Buck, former Director of the Department of Agricultural Economics, Nanking University, organized and completed two large-scale rural surveys during 1921–1925 and 1929–1934, and published two monographs, i.e., *Chinese Farm Economy* and *Land Utilization in China*. After several years of work, the College of Economics and Management, Nanjing Agricultural University (NJAU) successfully sorted out and restored the whole set of original data of Buck's 1929–1934 survey stored in the university and obtained detailed figures in respect of agricultural production and rural living, which covered 16,786 sample households and 168 cities and counties across 22 provinces of China.

[3] In *Land Utilization in China*, Buck divided China's agriculture regions into Wheat Zones and Rice Zones. The former includes the Spring Wheat Area, Winter Wheat–Millet Area, and Winter Wheat–Kaoliang Area. The latter include the Yangtze Rice–Wheat Area, Rice–Tea Area, Sichuan Rice Area, Double-Cropping Rice Area, and Southwestern Rice Area.

[4] In the second survey, Buck selected two groups of survey samples for Gaolan, identified as Gaolan (1) and Gaolan (2), respectively.

(Ningsia), and one locality in Tsinghai (Hwangyuan). These data recorded farm households' tenancy, land usage, cropping structure, tenancy system, off-farm employment, and hired labor. Although these sample counties and households may not reflect the complete picture of tenancy relationships in Northwest China, these survey data are, so far, the only available and reliable data at the household level for the Republican era.[5] With this caveat in mind, our analyses of these 610 households in six localities can at least provide a glimpse of tenancy and land rental arrangements in modern Northwest China.

6.2 An Overview of Tenancy in Northwest China

Tenant farming refers to those farm households that rent-in land, including landless tenants and part-owners. Landless tenants completely relied on rented-in land, whereas part-owners owned some of their landholdings. With this concept of "owned landholdings" we can measure the division of landless tenant farmer and self-owner farmer and estimate the percentage of land rented. However, this measure is not perfect and some subjectivity is required. For example, from Buck's data the No. 9 farm household in Kaolan (1) had both owned land and rented-in land, among which all the owned land (0.103 mou) was used for constructing farmhouses and tombs and the rented-in land (1.545 mou) was the sole land used for crop production. If consideration is only given to the ratio of owned land to total land, then various statistics and survey research might differ in the identification of such farm households. This chapter holds that it would be more accurate to define these farm households as landless tenant farmers as far as they were concerned with crop production. So landless tenant farmers in this chapter refer to farm households who rented in all of their land for crop production (cropland) only.[6]

As shown in Table 6.1, in the 1930s owner farm households were dominant in six localities in three provinces of Northwest China. Ownership–tenancy proportions differed. For example, according to these data there were no landless tenant farmers at all in Wuwei and Ningsia. Although the

[5] As calculated by the authors on the basis of Buck's survey data, the per capita cropland in the southern region and the northern region of China in the 1920s and 1930s was 2.28 mou and 3.34 mou, respectively, which accords with the estimations of per capita cropland by Zhang Youyi (1991) at approximately 2.34 mou in the southern region and 3.56 mou in the northern region.

[6] The term "land" used alone here and in what follows means all land irrespective of usage. "Cropland" specifically means that part of land used for agricultural production.

Table 6.1 Types of farm households (%)

Provinces and counties	*Percentage of types of farm households*		
	Owner farmers	*Part-owner farmers*	*Landless tenant farmers*
Kansu, Kaolan (1)	86.00	11.00	3.00
Kaolan (2)	67.00	18.00	15.00
Wuwei	96.00	4.00	0.00
Tianshui	62.00	21.00	17.00
Ningsia, Ningsia	98.00	2.00	0.00
Qinghai, Hwangyuan	67.27	22.73	10.00

Source: Adapted by the authors from Buck's second rural survey for China

county samples are limited, Buck's survey data appears to confirm the existing academic judgments and conventional wisdom on land issues of the northwest that there were fewer landlords in Northwest China than the Northern Plain or South China, that the degree of concentration of landholdings was far less than that of the national average level,[7] and that the tenancy economy in Northwest China was, as a whole, backward.

However, the composition of tenant farmers was very complex. Research on other regions of modern China revealed that among tenant farmers there were rich farmers similar to managerial farmers in capitalist societies as well as poor farmers who could barely afford food and shelter. The traditional view of equating the welfare and living conditions of tenant farmers as being equivalent to small land-owning farmers may not be accurate. By conducting in-depth analyses on tenancy behaviors of various types of tenant farmers from a micro perspective in Northwest China during the Republican era, we provide a data-rich supplement to existing studies that should lead to more clarification and understanding.

6.3 Causes of Land Renting and Production Behaviors of Tenant Farmers in Northwest China

Previous studies, e.g. Kung et al. (2012), usually attribute land rental by farm households primarily to the natural conflict between land and people.[8] This conflict arises from a number of different sources, but primarily the

[7] Liu Kexiang (2001).

[8] As pointed out by Kung et al. (2012), one of the most significant causes for land tenancy behaviors in modern China was the contradiction between land and people, arising from limited cropland resources.

cause is equivalent to a Malthusian trap, in the sense that the population growth rate increases at a rate greater than that at which new land can be brought into production; that new lands are of lower quality than historically occupied lands; and that local dynasties that emerged from earlier settlement had a prior advantage on landholdings per capita over newly established or newly emigrated farming households (Fu and Turvey 2018).

When the conflict between land and people is measured by "per capita owned cropland area," we identified two extremes of part-owner farmers in Northwest China. One extreme was in areas where cropland resources were scarce, and the per capita cropland area was far less than the regional average. The other extreme was that in areas where cropland was relatively abundant the per capita cropland area was far more than the regional average. Those farm households with per capita cropland area around the regional average value were unexceptionally self-owner farmers. For the sake of simplicity, part-owner farmers in these two circumstances are expressed as "small part-owner farmer" and "big part-owner farmer." As shown in Table 6.2, the various regions in Northwest China differed in per capita cropland area, and although there was a small number of part-owner farmers in Ningsia, there generally existed both types of part-owner farmers in all counties to different extents. As observed from the original data, small part-owner farmers were confronted with a prominent conflict between land and people, while big part-owner farmers owned abundant land resources and some of them even rented in land while leaving part of their owned land idle.

Table 6.2 Analysis on land and people contradiction confronted by part-owner farmers in Northwest China (unit: mou)

Provinces and counties	*Per capita area of owned cropland*		*Regional per capita area of cropland*
	Small part-owner farmers	*Big part-owner farmers*	
Kansu, Kaolan (1)	0.68 (81.82%)	3.56 (18.18%)	1.58
Kaolan (2)	0.61 (88.89%)	4.27 (11.11%)	1.86
Wuwei	2.29 (75.00%)	8.78 (25.00%)	5.52
Tianshui	1.58 (85.71%)	5.37 (14.29%)	3.18
Ningsia, Ningsia	1.99 (100%)	/	4.17
Qinghai, Hwangyuan	3.56 (84.00%)	12.48 (16.00%)	7.79

Source: Adapted by the authors from Buck's second rural survey for China

Note: Values in parentheses are the percentage of a particular type in relation to all part-owner farmers

6.4 "Subsistence-Type" Tenant Farmers

As discussed, the conflict between land and people was most prominent for landless tenant farmers, who had to rely on renting land to generate income. Meanwhile, those small part-owner farmers with little owned land were confronted with similar sharp contradictions. It is worth noting that the culture in Northwest China was (and still is) to bury the dead on farmlands; the spiritual bond of doing so was so great that even farm households with little cropland felt obligated to follow this tradition. In extreme cases, some farm households used up to two thirds of their owned land for building tombs, thus further intensifying the shortage of cropland for cultivation.

Landless tenant farmers and small part-owner farmers constituted the group most affected by land/population pressures and trapped in persistent poverty. Their only hope of breaking out of this poverty trap was to rent-in land. Though their form of landholdings differed, these farmers were persistently poor and of the "subsistence type" of tenant farmers. As shown in Table 6.3, the cropland area per household of landless tenant farmers in all regions was low. Conditions for small part-owner farmers were a little better, but their cropland area per household was still lower than the regional average. As revealed by existing studies, rich farmers could rent-in land easier than ordinary farmers, regardless of the tenancy system development level.[9] If wealth was indicated by owned land, it was, naturally, more difficult for landless tenant farmers and small part-owner farmers with little land resources to obtain sufficient land.

"Subsistence-type" tenant farmers had very small areas of cropland and could generally rely on family labor to perform farm work.[10] They used most of the rented-in land for growing grain crops such as wheat, millet, kaoliang, and soybean to feed their own families, but many also had to rely on off-farm employment to maintain subsistence levels. As presented in Table 6.4, a significant percentage of family members were employed in off-farm labor and this was an important source of income. In particular, for landless tenant farmers, off-farm income accounted for over half of their total family income. As observed from Buck's original survey data, off-farm employment in Northwest China included a variety of forms, such as hired long-term labor, seasonal labor, or working as a blacksmith, stonemason, peddler, weaver, or servant.

[9] Shi Jianyun (1998).

[10] As calculated according to Buck's original household data, for small part-owner peasants and landless tenant peasants in Northwest China family accounted for nearly 100% of all laborers.

Table 6.3 Cropland area per household for "subsistence-type" tenants

Provinces and counties	*Small part-owner farmers*			*Landless tenant farmers*	*Regional average*
	Owned area	*Rented area*	*Total*		
Gansu, Kaolan (1)	4.54	3.42	7.96	3.08	11.21
Kaolan (2)	4.64	3.27	7.91	4.38	12.39
Wuwei	21.67	6.85	28.52	/	36.64
Tianshui	8.20	5.21	13.41	11.52	15.87
Ningsia, Ningsia	10.74	6.28	17.02	/	22.31
Qinghai, Hwangyuan	20.25	22.11	42.36	39.37	68.70

Source: Adapted the authors from Buck's second rural survey for China

Table 6.4 Distribution and income of off-farm employment of "subsistence-type" tenant farmers in Northwest China

Provinces and counties	*Small part-owner farmers*		*Landless tenant farmers*	
	Percentage of off-farm hired labor in family members (%)	*Percentage of off-farm income in total income (%)*	*Percentage of off-farm hired labor in family members (%)*	*Percentage of off-farm income in total income (%)*
Kansu, Kaolan (1)	14.28	46.52	37.23	66.50
Kaolan (2)	5.68	22.17	19.28	59.62
Wuwei	37.74	22.59	/	
Tianshui	19.55	52.26	22.92	52.44
Ningsia, Ningsia	12.25	11.78	/	
Qinghai, Hwangyuan	8.26	19.25	15.17	45.00

Source: Adapted the authors from Buck's second rural survey for China

Note: In order to eliminate the influence of factors such as gender and age on the actual quality of laborers, Buck converted all laborers into adult equivalent laborers in the following criteria: one man = 1 equivalent adult laborer, one woman = 0.8 equivalent adult laborer, one child = 0.5 equivalent adult laborer. Labor data here have been transformed in such a way

6.5 Cropping Structure: "Improving Living Condition-Type" Tenant Farmers

Beside the great number of "subsistence-type" tenant farmers, there were also about 10% big part-owner farmers who had abundant resources of owned land and even owned idle land. The reason for this type of tenant farmer to rent-in land is obviously unlinked with the contradiction between land and people. Based on the sorted original data of Buck's survey, the authors hold that this may be attributed to the requirements of the cropping structure, that is, special requirements of some crop varieties for particular soil quality. Specifically, some of the owned land of big part-owner farmers in Kaolan (1), Wuwei, Tianshui, and Hwangyuan, due to poor irrigation conditions, could only be used for growing coarse cereals such as millet rather than fine grain such as wheat with a better taste and higher nutritional value. In addition, due to the lower marketing rate of agricultural products in these areas, farmers rented in land in valleys or mountains with better irrigation conditions to plant wheat to supply their own families and improve their diet structure and living quality. The case for Kaolan (2) region was special, as there were many military depots set on army land, which accounted for 43.6% of total land in this region,[11] so there was a higher demand for opium and tobacco. A number of farmers in this region left their owned land idle, and, instead, rented in appropriate land to plant and sell opium and tobacco for cash.

These big part-owner farmers owned cropland sufficient to maintain or even exceed the demand for family subsistence. The cause of their land-renting behavior, intended either to improve the family diet structure or to obtain more cash income, was the demand for special crops. Such land renting was a kind of tenancy for improving living conditions. As shown in Table 6.5, these farmers rented in more land than the two lower categories of subsistence farmers. Here we observe the "Mathew effect" in the tenancy market of modern Northwest China, where the area of land rented in by tenant farmers increased with the area of owned land, perhaps crowding out the opportunity for tenancy by poorer, more limited-resource households.

Large part-owner farmers, with a larger area of cropland, had to rely on hired labor in varying degrees to perform over 20% of the farm work. In addition, this type of tenant relied less on off-farm employment. As shown in Table 6.6, except for Tianshui, off-farm employment of these tenant farmers was lower in terms of family members and contribution to income. In Wuwei and Hwangyuan, there were no records of off-farm employment at all.

[11] In modern China, most land was privately owned, except for a certain area of officially owned cropland, school-owned cropland, temple-owned cropland, clans-owned cropland, army land, and relief cropland in a few regions.

Table 6.5 Cropland area per households of "improving living conditions-type" above-subsistence tenant farmers in Northwest China (unit: mou)

Provinces and counties	*Big part-owner farmers*		
	Owned cropland area	*Rented in cropland area*	*Total*
Kansu, Kaolan (1)	17.95	11.11	29.06
Kaolan (2)	43.65	9.16	52.81
Wuwei	37.20	9.43	46.63
Tianshui	20.98	36.71	57.69
Ningsia, Ningsia	/		
Qinghai, Hwangyuan	87.02	38.38	125.40

Source: Adapted by the authors from Buck's second rural survey for China

Table 6.6 Labor distribution and income of off-farm employment for "improving living conditions-type" tenant farmers in Northwest China

Provinces and counties	*Big part-owner farmers*	
	Percentage of off-farm hired labor in family members (%)	*Percentage of off-farm income in total income (%)*
Kansu, Kaolan (1)	1.12	3.33
Kaolan (2)	4.04	12.00
Wuwei	0	0
Tianshui	13.52	40.20
Ningsia, Ningsia	/	
Qinghai, Hwangyuan	0	0

Source: Adapted by the authors from Buck's second rural survey for China

Note: The labor figures have been converted to adult equivalent labor

On the other hand, and as observed in Buck's original survey data, the higher off-farm employment percentage in Tianshui might be linked with the characteristics of the local rural economy, where men were engaged in selling firewood, seasonal jobs, and handcraft work, and nearly all the women and children in weaving and spinning.

Reexamining Buck's data, we find that the nature of tenancy in the Republican era was complex. Our examination of the data reveals that landless tenant farmers and small part-owner farmers rented in land because of the pressures between land and people. Tenancy was a necessity if these households sought to escape the poverty trap and survive at even the most rudimentary levels of subsistence. In addition to renting land, many households relied on off-farm employment to maintain subsistence. For larger tenancy arrangements, there was a greater reliance on hiring-in labor and less need for household members to seek off-farm employment.

6.6 Tenancy System in Northwest China

Similar to that in modern North China,[12] the land tenancy system in Northwest China in the same period included share rent, cash crop rent, and cash rent.[13] For share rent, the output of harvest was shared by landlord and tenant in a certain proportion and the rental rate was directly expressed in share proportion (%). For the latter two forms of rental arrangements, the quantity of cereal or currency to be paid as rental was determined before the land was rented out. As observed in Buck's original data, in Northwest China landlords interfered very little in a tenant's operation of land. However, some restrictions might well have been applied. For example, in reviewing the record for farm No. 99 in Kaolan (2), it was noted that "the land shall be returned to the landlord if not properly operated by the tenant."

Tenancy systems were quite different and complex. After sorting through Buck's original survey data, we found that differences existed not only in tenancy systems across different regions, but also in tenancy systems and rental rates for different types of tenant farmers. Table 6.7 presents the characteristics of tenancy systems in Northwest China. Except in Wuwei where cash crop rent was prevalent, share rent was prevalent, particularly in Kaolan (2) and Tianshui. Research has shown that in land rental markets without an insurance system, the preferred method of mitigating crop production risks was through a share rent system between landlords and tenants.[14] Though the risk preference of landlords and tenants in Northwest

[12] Shi Jianyun (1997).

[13] All these rental forms are named in Buck's *Land Utilization in China*. Share rent is the division of crops in a certain proportion such as 40% to the landlord and 60% to the tenant. Cash crop rent is the payment of a definite amount of grain by the tenant to the landlord. Cash rent is the payment of a definite amount of money by the tenant for the use of the landlord's land. Buck also mentioned in his book another rental form that did not exist in Northwest China—cropper, which is when the tenant supplies chiefly labor in return for a certain proportion of the crop; this proportion is always smaller than in the share rent system. As this chapter is based on Buck's original household data, Buck's naming method is adopted here.

[14] J. G. Sutinen, (1975).

Table 6.7 Characteristics of tenancy systems in Northwest China

Provinces and counties	*Tenancy system*	*Tenancy system for different types of tenant farmers*			
		Tenant farmers renting in land for subsistence		*Tenant farmers renting in land to improve living conditions*	
		Tenancy system	*Rental rate*	*Tenancy system*	*Rental rate*
Kansu, Kaolan (1)	Share rent, cash rent, and cash crop rent coexisted	Share rent Cash rent	Unknown	Cash crop rent	Unknown
Kaolan (2)	Mostly share rent, some cash rent	Share rent	50%	Share rent Cash rent	40%
Wuwei	Mostly cash crop rent, some cash rent	Cash crop rent	30%	Cash crop rent Cash rent	15%
Tianshui	Mostly share rent, some cash rent	Share rent Cash rent	50%	Share rent	25%
Ningsia, Ningsia	Share rent and cash rent coexisted	Share rent Cash rent	60%	/	
Qinghai, Hwangyuan	Share rent and cash crop rent coexisted	Shar e rent	50%	Cash crop rent	65%

Source: Adapted by the authors from Buck's second rural survey for China

Note: Buck's original data had no record of cash rent, so data on rental rate under the arrangement of cash rent are missing in this table

China was unknown, share rent was the most common form of land leasing arrangement in that area. As can be observed in Table 6.7, share rent was dominant in Kaolan (2) and Tianshui, which had the highest number of tenant farmers and an active tenancy economy. However, cash crop rent was dominant in the Wuwei district where there were the fewest number of tenant farmers. Various forms of land leasing coexisted in other areas.

It is worth noting that the tenancy systems and rental rates differed between "subsistence-type" tenant farmers and "improving living conditions-type" tenant farmers. Next we explore the characteristics of the three types of tenancy systems and the possible causes that led to differentiation of the tenancy systems and rental rates in Northwest China.

6.6.1 Production Responsibilities

As shown in Table 6.7, in Kaolan (2), Tianshui, and Wuwei there was one tenancy system that was basically the same for tenant farmers renting in land for subsistence and those who hoped to improve their living conditions; although the former, obviously, had to pay higher rental rates. The authors hold that in Tianshui and Kaolan (2), where share rent was prevalent, the difference in rental rates might be linked to production responsibilities.

Share rent is a tenancy arrangement under which the landlord and tenant share production responsibilities and outputs as well. Therefore, the share proportion is usually linked to some extent to the production responsibilities shared by the parties and the varieties of crop.[15] Under share rent in Northwest China, tenants had to pay rent via a certain proportion of all outputs harvested on the rented land. For example, if the tenant sowed wheat in spring and planted corn or millet after harvesting wheat in summer, the outputs of both crops would be shared with the landlord. Of course, there were some exceptions, for example, a few tenants increased the proportion of, or even gave all, opium output as rent, or paid rent with opium planted on their owned land so as to cut down the rental rate for cereal crops such as wheat and millet planted on rented land. In Kaolan (2) and Tianshui, the rental rates were some 50% for tenant farmers renting in land for subsistence, and only 25–40% for those hoping to improve their living conditions. Decreasing rental rates with the increase of rented-in land area might be linked with the better operation abilities of those tenant farmers who rented in land to improve their living conditions. This type of tenant farmer could input production factors, such as seeds and tools, themselves, and thus the landlord assumed less production responsibilities and accordingly shared less percentage of the output.

In Kaolan (1), Ningsia, and Hwangyuan, share rent and other forms of tenancy coexisted. Share rent was dominant when tenant farmers rented in land for subsistence. This was likely linked with production responsibilities too. These tenant farmers were relatively poor and could barely acquire the inputs necessary for production, so they had to select share rent to obtain the support of input from the landlord. Except for Kaolan (1) where the rental rate was unknown, outputs were shared between landlord and tenant at a proportion of 40:60% (40% for tenant and 60% for landlord) in Ningsia and around 50% in Hwangyuan.

[15] Shi Jianyun (1998).

6.6.2 Negotiation Ability

In Wuwei, where a cash crop rental system was common, tenant farmers who rented in land to improve their living conditions enjoyed advantageous rates. However, the estimation of rental rates under the cereal rental arrangement is very complex and requires ascertaining the crop varieties and amount per mou paid by tenants. As there were no uniform units of measurement across different regions of China at that time, comparison is feasible only after data were converted to uniform units.[16] In addition, due to differences in land fertility and yield, the rental per mou may not accurately reflect the rental rates. Fortunately, in the statistics for crop yields, beside per mou yields of the year, Professor Buck also recorded in detail the "most frequent yield," "normal yield," and "best yield" of each sample farm household. These represented the "yield most often obtained within 10 years," "yield obtained under all conditions favorable for the crop," and "yield obtain[ed] under optimum conditions," respectively. The "most frequent yield" likely smooths differences in per mou yield across good and bad years caused by objective factors such as land quality, so the ratio of per mou rent to most frequent yield may more accurately and reasonably reflect the rental rates.

In Wuwei, most part-owner tenants paid rent under a cereal rent arrangement, usually with wheat as rent-in-kind. Observed only from rental amount per mou, the rental rates were different among tenant farmers, from 1 dou per local mou to 1.5 dou per local mou (1 dou per local

[16]The standard of weights and measures, mou system and measuring vessels at that time differed greatly across different regions of China. In Buck's original data, units of output, area, distance, and currency were quite complex. For example, units of output were expressed in jin (斤), dou (斗), dan (担), and dàn (石), unit of area was local mou, distance was expressed by an unnamed local unit, currency units included Diao, Yuan, and Silver Yuan (Yin Yuan). These units not only differ from the metric system but also across regions. In other words, jin in different places does not equal *shi jin* and is converted to the metric system at different ratios. During the process of sorting Buck's original data (see Chap. 4), the research group obtained relevant figures of output and areas by referring to the conversion ratios between jin in different regions and the metric kilogram, and to that between local mou and *shi mou*/hectare, as recorded in detail by Buck in his book *Land Utilization in China*. Meanwhile, through calculation and comparison of the mean values recorded in the original data for different regions with the mean values that had been converted to metric units in *Land Utilization in China*, the research group obtained the conversion ratios between jin, dou, dan, and dàn for different crops and the metric unit kilogram, the conversion ratios between local unit of distance and metric unit meter, and the corresponding currency conversion ratios.

mou is equivalent to approximately 37.03 jin/mou). Further calculation of the ratio of per mou rent to most frequent yield revealed that the rental rates ranged from 16.67% to 37.50% of most frequent yields and were lower for tenant farmers who rented land to improve their living conditions. This might be explained by their advantage in owned land resources which granted them a favorable position in the negotiation with landlords over rental rates.

6.6.3 Risk-Resistance Capacity and Rent Sensitivity

In Kaolan (1) and Hwangyuan, where cash crop rent and other tenancy systems coexisted, tenant farmers who rented in land to improve their living conditions mostly preferred to cash crop rent. This might be explained by their higher risk-resistance capacity.

Rental rates in Kaolan (1) are unknown. The situation of rent-in-kind in Hwangyuan was quite complex. Some tenants paid rent with barley or horse bean once a year, while others paid rent twice a year with barley/broad bean and soybean/wheat. The per mou rental rates calculated with regard to crop products varied due to the difference in price of crop products, but, basically, the amount of rent paid once a year was approximately twice that paid twice a year. Moreover, as the market price of wheat was higher, the rental rate expressed with wheat (1 dou/local mou) was relatively lower (1 dou/local mou equivalent to 80.01 jin/mou) than those expressed in other crop products, which ranged from 2 to 2.5 dou/local mou. As observed, simply from the absolute value of per mou rent, the rent burden for tenant farmers in Hwangyuan was obviously heavier than that for those in Wuwei. Nevertheless, due to possible differences in yield level between these two areas, it is necessary to further compare the ratio of rent amount to most frequent yield.[17] Rent-in-kind accounted for around a third of output where rent was paid twice a year, and about 50% and 70% if paid in barley and horse bean, respectively, where rent was paid once a year. Obviously, the rent burden for tenants in Hwangyuan area was much heavier compared with that in Wuwei. The authors consider that this might be explained by the lower rent sensitivity of tenants in Hwangyuan due to relatively more abundant land resources and lower pressure of subsistence.

[17] Buck defined the estimation duration for usual yield as ten years, so for all peasants who grew crops for less than ten years, the data were recorded as "U" (unknown). The usual yields of a few tenant peasants in Huangyuan was unknown and were substituted for by the per mou yield of the year.

6.6.4 *Pressure of Subsistence and Crop Varieties*

As presented in Table 6.7, in some areas, cash rent was prevalent among "subsistence-type" tenant farmers, while in other areas it was prevalent among tenant farmers who rented in land to improve their living conditions. Though the rental rates under the cash rent system are unknown, the fact this system was used for such diverse cases might be related to the pressure of subsistence and crop varieties.

It is generally considered that cash rent is closely related to a commodity economy and adopted for land which is used to grow economic crops or wheat.[18] This is confirmed by the use of cash rent in Kaolan (2) and Wuwei, where nearly all tenants paying cash rent were those who: (a) rented in land to improve living conditions; (b) lived above subsistence level so were not under this pressure; (c) were involved in no off-farm employment; (d) grew higher value wheat, tobacco, and opium on rented-in land; and (e) sold 20% of their wheat and almost all tobacco and opium to obtain a cash income.

The case in Gaoalan (1), Tianshui, and Ningsia, however, was quite different, as cash rent was only widely adopted here by "subsistence-type" tenant farmers. Facing greater pressure, this type of farmer grew coarse cereals such as millet,[19] instead of economic crops. This would provide a sufficient food supply targeted to meet the demand of subsistence to the maximum extent under the environment of an undeveloped commodity economy. All crop outputs, instead of being sold, were used to feed the family.[20] In Tianshui, cash rent was only adopted by landless tenant farmers, who rented in land to grow coarse cereals such as kaoliang, corn, and millet to feed the family. On the premise that Buck omitted no data on sale of crop outputs, the money for cash rent might come from off-farm employment. These "subsistence-type" tenants used their limited area of cropland to grow low-value coarse cereals, which were not likely to make

[18] Shi Jianyun (1997).

[19] Based on the original data of Buck's survey, we find that in Northwest China where commodity economy was not well developed, small peasants (whether they rented-in land or not) preferred to plant coarse cereal crops. On the one hand, this decision was likely to be linked to the quality of their land, and, on the other hand, the higher unit output and calories of coarse cereals could more easily meet family demand for foodstuffs.

[20] As mentioned, in Buck's original household data, the quantity of output of each crop sold by each sample farm household was recorded in detail under "Quantity of Outputs for Various Usages by Crops." The authors find that no tenant peasants under the cash rent arrangement sold their crop output—it was all consumed by their family.

enough to pay land rent after much was used to feed their families. Therefore, they had to transfer surplus family labor to off-farm employment and earn off-farm income to pay land rent while balancing the pressures between land and people.

6.7 Conclusions

In this chapter we have shown that tenancy in Northwest China during the Republican era was not well developed on the whole. Most of the very few cases of land leasing occurred due to the pressures between land and people. A portion of large part-owner farmers with rich land resources leased in and operated land to meet the requirements of cropping structures. In other instances land leasing was used for subsistence-type farming, characterized by limited land area, production work performed mostly by family members, and heavy reliance on off-farm employment for subsistence. As for the tenancy system, share rent, cash rent, and cash crop rent were adopted in Northwest China to different extents, with share rent dominating on the whole in proportion to the number of tenant farmers in the area. In addition, influenced by production responsibilities, negotiation ability, risk-resistance capacity, rent sensitivity, pressure of subsistence, and crop varieties, the tenancy system and rental burden for tenants under these two types of leasing were different.

Of course, this study is not the end point for research on issues of tenancy and the tremendous and complex original data of Buck's survey, which require more effort from academic circles. Further research in more detail and across more regions is required. Moreover, a comparison study of Buck's data with other historical data for the same period may provide more definitive answers to issues pertaining to agriculture and the rural economy of modern China.

References

Buck, J.L. 1930. *Chinese Farm Economy: A Study of 2866 Farms in Seventeen Localities and Seven Provinces in China*. Chicago: University of Chicago Press.

———. 1937a. *Land Utilization in China: A Study of 16,786 Farms I 168 Localities, and 38,256 Farm Families in Twenty-Two Provinces in China, 1929–1933*. University of Nanking (1937), University of Chicago Press, Chicago (1937), and The Council on Economic and Cultural Affairs, NY NT, 1956.

———. 1937b. *Land Utilization in China-Atlas*. Chicago: University of Nanking (1937), University of Chicago Press.

———. 1937c. *Land Utilization in China-Statistics*. Chicago: University of Nanking (1937), University of Chicago Press.

Fu, Hong, and Calum G. Turvey. 2018. *The Evolution of Agricultural Credit During China's Republican Era, 1912–1949*. Cham: Palgrave Macmillan.

Kung, James Kai-sing, Xiaogang Wu, and Yuxiao Wu. 2012. Inequality of Land Tenure and Revolutionary Outcome: An Economic Analysis of China's Land Reform of 1946–1952. *Explorations in Economic History* 49 (4): 482–497.

Li, Jinzhen. 2011. No Overcorrecting: Complex Reality of Tenancy Relationship in Modern North China Plain Viewed from Jizhong County. *Modern Chinese History Studies* 6: 124–135.

Liu, Kexiang. 2001. General Investigation and Quantity Estimation of Land Classes Distribution in the 1930s of China—Studies on Land Issues in the 1930s (2). *Researches in Chinese Economic History* 3: 19–35.

Shi, Jianyun. 1997. Study on Land Rental Forms in Modern North China Plain—Exploration of Tenancy Relations in Modern North China Plain. *Modern Chinese History Studies* 3: 168–184.

———. 1998. Land Operation and Rental Burden in Modern North China Plain—Exploration of Tenancy Relations in Modern North China Plain. *Modern Chinese History Studies* 6: 99–119.

Sutinen, J.G. 1975. The Rational Choice of Share Rent and Implications for Efficiency. *American Journal of Agricultural Economics* 57 (4): 613–621.

Zhang, Youyi. 1991. Re-Estimation of Population and Cropland in Modern China. *Researches on Chinese Economic History* 1: 20–30.

CHAPTER 7

Regional Differences in Surplus Agricultural Labor During China's Republican Era, Based on Buck's Rural Survey Data

Hao Hu and Weiwei Zheng

7.1 Introduction

In this chapter we explore the problem of surplus agricultural labor in the mid-Republican era. With a large population and little cropland, the high population-to-land ratio has been recognized as an impediment to economic growth and greater-than-subsistence levels in rural China since ancient times. Scholars conducting research on the quantity of surplus labor have come to varied conclusions. For example, Funing Zhong holds that such difference is mostly attributed to "the inconsistency in statistics of rural and agricultural labor forces."[1] The "rural" labor force and "agricultural"

[1] Zhong Funing, (1995).

H. Hu (✉)
College of Economics and Management, Nanjing Agricultural University, Nanjing, China
e-mail: huhao@njau.edu.cn

W. Zheng
Jiangsu Academy of Agricultural Sciences, Nanjing, China

H. Hu et al. (eds.), *Chinese Agriculture in the 1930s*,
https://doi.org/10.1007/978-3-030-12688-9_7

labor force are two completely different concepts. The former refers to laborers who reside in rural areas, including those engaged in agricultural, secondary, and tertiary sectors. The latter refers specifically to laborers engaged in agricultural production, with relevance to occupation. The degree of consistency between these two concepts depends on city division standards and the degree of non-agriculturization of rural areas. During the Republican era, the secondary and tertiary sectors of China's rural economy lagged behind more urban regions, and far behind more developed economies. With the degree of non-agriculturization of rural areas being quite low, however, rural labor could, for the most part, be regarded as agricultural labor, or at least proportionately so. Therefore, it is feasible to estimate the agricultural labor surplus in the Republican era on the basis of rural labor figures from Buck's 1929–1933 survey because all respondents were in fact farm households. Since non-farm businesses such as shops, restaurants, teahouses, and small manufacturing units made up such a small part of the rural economy, we can speak generally about labor surplus and agriculture, without degrading the concept of rural labor in a meaningful way.

The existing literature on surplus agricultural labor during the era of the Republic of China was restricted by the availability of data and provided only rough estimates. Li Zhang,[2] using survey data from Man Tie San Cun Village, estimates that the percentage of surplus men labor in rural Wuxi was 65%, based on the assumption that "*all farm works were performed by men and housework and silkworm breeding by women*" and on the basis of empirical ratio of rice planting area to mulberry planting area (80% to 20%). Jiafu Han concluded on empirical statistics that "*the number of days each year suitable for cropping was no more than 100 for regions in the vicinity of Xinganling Mountain, Western Manchuria, 200 for North China Plain and 350 for Sichuan Basin,*" and observed that that "*job opportunities for peasants aged between 15 to 45 of the whole country lost in idle season were equivalent to 55,000,000 peasants completely unemployed.*"[3] Muqiao Xue,[4] using a survey on rural Wuxi, that on average poor peasants undertook agricultural work for 148 days each year, middle peasants 152 days, and rich peasants 181 days, suggesting that on average able-bodied men engaged in agricultural work were idle for more than a half of

[2] Zhang Li, (2007).
[3] Han Jiafu, (1945).
[4] Xue Muqiao, (1946).

each year. However, in practice empirical estimates such as these might be biased since farm households differed in the labor distribution between men and women, and these estimates did not fully account for different cropping patterns and labor requirements across regions.

Zhongjian Zhou is perhaps the earliest scholar to deviate from this estimation method.[5] Starting his analysis from the average number of people and labor per households, he made theoretic calculations on the supply and demand of agricultural labor in southern Jiangsu from the perspectives of the cropping abilities of agricultural labor and the per mou labor requirements of major crops. He compared the calculation results with field survey data at that time and concluded that the potential surplus rate of agricultural labor in southern Jiangsu was around 40% and the seasonal surplus rate was above 70%. Zhou's research is based on assumptions that "*rural laborers were only engaged in farming*," "*surplus rural labor was surplus agricultural labor*," "*rural labor productivity remained unchanged*," and "*the net outflow rate of rural labor was zero*." Because survey data, including Buck's, is generally of a short-term nature these assumptions are not unreasonable. However, Zhou based his conclusion that "*there was significant difference in surplus agricultural labor*" on his estimation of surplus labor of southern Jiangsu only.

In this chapter we assess the issue of surplus labor across different agricultural regions of China and make comparisons across different cropping patterns, so as to provide more reliable and fundamental figures for further studies. In particular, we draw on Buck's data to estimate the surplus agricultural labor in the six farming areas of the rice zone and wheat zone using estimation methods from classical economics, and then consider factors of influence such as endowments of interregional agricultural production resources, cropping structures, and farming patterns.

7.2 Methods for Estimation of Surplus Agricultural Labor

Methods for estimating surplus agricultural labor can be put into three categories. The first is estimation methodology of classical economics. According to the principles of classical economics, land and capital are relatively scarce and labor supply is unlimited. Lewis proposed "*that if a certain quantity of labor is extracted from traditional sectors without reducing the*

[5] Zhou Zhongjian, (1997).

total amount of output, then this part of labor is surplus labor, i.e., *the part of labor with zero marginal productivity*."[6] This category has two estimation methodologies. One is estimation of surplus agricultural labor in a narrow sense: "*the difference between the quantity of available agricultural labor and the quantity of demand of labor by agriculture under existing agricultural production technology and farming methods*." The other is estimation of surplus agricultural labor in a broad sense: "*the difference between the quantity of available agricultural labor and the quantity of demand of labor by agriculture under advanced agricultural production and management technology*." In comparison, because it is influenced by too many variable factors, surplus labor in the broad sense does not really contribute to an understanding of actual surplus labor.

The second category is in the estimation methodology of neoclassical economics. The school of neoclassical economics holds that Lewis's surplus labor with zero marginal productivity does not exist,[7] but surplus labor with marginal productivity greater than zero probably does.[8] The common estimation method is to obtain the quantity of surplus labor by deducting the quantity of labor actually existing in agriculture from the quantity of agricultural labor required for maximization of benefits of the whole economic system (i.e. optimum allocation of resources). This estimation methodology considers labor allocation simply from the market perspective, without taking into account agricultural technology level and peasants' farming conditions. In the early twentieth century, the commodity economy of China's agricultural products was fairly new and the market capacity was not strong enough to maximize resources allocation, so estimation methodologies of neoclassical economics are not suitable for estimating the surplus labor in that period.

The third approach is a standard structure comparison estimation method. Here the output value and employment proportions of industrial sectors correspond to per capital GDP levels measured relative to international standards. In other words, the quantity of surplus labor of China would be obtained by comparing the total quantity of labor in China's agricultural sector with the corresponding international standard for the same GDP level. Surplus agricultural labor calculated in this method is surplus agricultural labor in a narrow sense. However, due to lack of systematic statistical GDP data for early twentieth-century China, this standard structure

[6] W.A. Lewis, (1989).
[7] Theodore W. Schultz, (1999).
[8] W. Jorgenson, (1967).

comparison estimation method cannot generally be applied to the estimation of surplus agricultural labor in that period.

From these various methodologies, and given the nature of the data available to us, this chapter uses the classical economics methodology for estimation of surplus labor in China during the early twentieth century, drawing on Buck's rural household survey data for China.

7.3 Data Sources and Research Methods

We make our estimates using a classical economics methodology for estimation of surplus labor in narrow sense and in combination with the Buck's survey data. More specifically we use:

Surplus agricultural labor = available agricultural labor − demand of labor by agriculture

Rate of surplus agricultural labor = (surplus agricultural labor/available agricultural) × 100%

There are two assumptions here. One is that the net inflow of rural labor of the sample county is assumed to be zero since the available rural labor within the sample counties is used as the supply quantity of agricultural labor. In 1930s China, rural laborers mostly worked in fields and during the fallow season might have done some subsidiary work. Due to limitations in transport and road infrastructure, laborers moved within a village or among villages in the vicinity. The net inflow of labor would have been negligible, at least in the short term for which our data holds. The second assumption is that agriculture is "small-scale farming," so that the demand for labor by agriculture is considered only in the context of labor demand by small-scale farms.

The quantity of available agricultural labor here is expressed as the quantity of agricultural labor timed by annual working days of unit labor. The quantity of agricultural labor is converted using the conversion standard set in *Chinese Farm Economy* (Buck 1930) for equivalent adult labor.[9] For annual working days of unit labor, this chapter refers to Zhou's definition on working time and sets 300 working days for each unit of labor.[10] For demand of labor by agriculture, in *Statistics Information*, Buck recorded in detail the quantity of labor required for each unit area of each crop in each region. In this chapter, it is expressed as:

[9] John L. Buck, (1930).

[10] Zhou Zhongjian, (1997).

demand of laborby agriculture = $\sum$ crop planting area × labor required for unit area of crop.

Planting areas of crops are recorded in detail in the statistic tables of Buck's rural survey.

7.4 Estimation Results and Analysis

Using this approach, the surplus agricultural labor of China in the Republican era is presented in Tables 7.1 and 7.2.

As presented in Table 7.1, the rate of surplus agricultural labor in the wheat zone was 79.43%, and over 60% of the regions were above the average surplus rates. The surplus rates were generally higher in regions of the rice zone. The surplus rate was the highest for the Winter Wheat–Millet Area (80.63%), less for the Winter Wheat–Kaoliang Area (78.87%), and lowest for the Spring Wheat Area (77.71%). Surplus rates differed significantly across regions within the same farming areas. For example, in the Spring Wheat Area, Dingbian, Shaanxi, had the highest surplus rate (85.87%) and Ningwu, Shanxi, had the lowest (66.58%). In the Winter Wheat–Millet Area, Jincheng, Shaanxi, had the highest surplus rate (88.97%), Fuping, Hebei, the lowest (62.69%). In the Winter Wheat–Kaoliang Area, Fuyang, Anhui, had the highest (88.47%), and Fushan, Shandong, the lowest (55.00%).

As presented in Table 7.2, the agricultural labor surplus rate in the rice zone was 69.22%, lower than that of the wheat zone. The surplus rate was higher for the Yangtze Rice–Wheat Area (70.64%) and the Rice–Tea Area (70.39%) and lower for the Southern Rice Area (66.63%). Surplus rates differed significantly across regions within the same farming areas of the rice zone. For example, in the Rice–Tea Area, Nanping, Fujian, had the highest surplus rate (88.10%) and Tangxi, Zhejiang, had the lowest (37.45%). In the Southern Rice Area, Mengzi, Yunnan, had the highest surplus rate (88.58%), Yiliang, Yunnan, had the lowest (34.03%). In the Yangtze Rice–Wheat Area, Wuxi 2, Jiangsu, had the highest (87.21%), and Yuyiao, Zhejiang, had the lowest (32.35%).

Analysis on data in Tables 7.1 and 7.2 shows that surplus labor in rural areas was tremendous in quantity and differed greatly across regions. The surplus rate was generally higher in the farming areas of the wheat zone than in the rice zone.

Table 7.1 Surplus agricultural labor

Year	*Region*	*Surplus rate (%)*	*Year*	*Region*	*Surplus rate (%)*
Spring Wheat Area [77.71%]					
1930	Dingbian, Shaanxi (100)	85.87	1931	Gaolan, Gansu 2 (100)	70.56
1931	Huangyuan, Qinghai (110)	79.33	1932	Wuwei, Gansu (100)	80.58
1931	Gaolan, Gansu 1 (100)	83.35	1932	Ningwu, Shanxi (100)	66.58
Winter Wheat–Millet Area [80.63%]					
1930	Linxian, Shanxi (100)	77.79	1931	Tianshui, Gansu (100)	85.51
1930	Jincheng, Shanxi (100)	88.97	1931	Zhenan, Shaanxi (98)	81.85
1931	Anyi, Shanxi (101)	76.44	1931	Zhouzhi, Shaanxi (100)	71.06
1931	Pingding, Shanxi (102)	83.75	1931	Weinan, Shaanxi (100)	81.79
1931	Fuping, Hebei (101)	62.69	1931	Xunyi, Shaanxi (100)	87.88
1931	Linbao, Henan (101)	75.62	1931	Mianxian, Shaanxi (100)	84.49
1931	Luoyang, Henan (102)	83.38	1932	Shangxian, Shaanxi (99)	85.05
1931	Pingliang, Gansu (100)	83.19			
Winter Wheat–Kaoliang Area [78.87%]					
1930	Changli, Hebei 2 (102)	84.99	1931	Tangyi, Shandong (102)	70.78
1930	Qinyang, Henan (90)	72.28	1931	Linzhang, Henan (100)	82.77
1931	Zhengding, Hebei (100)	82.04	1931	Shangqiu, Henan (100)	86.97
1931	Enxian, Shandong 2 (100)	88.37	1932	Fuyang, Anhui (99)	88.47
1931	Fushan, Shandong (100)	55.00	1932	Xiangcheng, Henan (100)	86.93
1931	Huimin, Shandong (101)	66.82	1932	Nanyang, Henan 1 (101)	83.13
1931	Laiyang, Shandong (100)	75.20	1933	Guanyun, Jiangsu (99)	80.48

Source: Calculations by the authors based on Buck's *China Land Utilization Statistics Information* and rural survey data

Note: Figures in parentheses are the quantity of farms, and figures in square brackets are agricultural labor surplus rates for the farming areas

In traditional farming, the surplus rate of agricultural labor is relevant, along with the endowments of agricultural production factors resources (especially area of cropland), cropping structure, and farming patterns. We explore below the major causes for regional difference in surplus agricultural labor in China from the ratio of cropland area owned by farm households to the area of rented-in land, crop structure, and production patterns.

Endowment of cropland resources and the proportion of land rented directly influenced labor input by farm households in different regions. In traditional agriculture, laborers in rural areas had to work on farms to earn

Table 7.2 Estimation results of surplus agricultural labor in the rice zone

Year	*Region*	*Surplus rate (%)*	*Year*	*Region*	*Surplus rate (%)*
Rice–Tea Area [70.39%]					
1931	Tangxi, Zhejiang (100)	37.45	1932	Xiuning, Anhui 1 (100)	67.18
1932	Linhai, Zhejiang (100)	70.19	1932	Xinhua, Hunan (100)	84.31
1932	Lishui, Zhejiang (100)	69.30	1932	Fuliang, Jiangxi (103)	69.60
1932	Tomnglu, Zhejiang 1 (105)	85.70	1932	Gao'an, Jiangxi (100)	79.27
1932	Tonglu, Zhejiang 2 (102)	70.68	1932	Duchang, Jiangxi (101)	45.26
1932	Yongjia, Zhejiang (100)	73.70	1933	Yiyang, Hunan (101)	74.34
1932	Nanping, Fujian (100)	88.10			
Southern Rice Area [66.63%]					
1932	Anshun, Guizhou (101)	55.31	1932	Mengzi, Yunnan (101)	88.58
1932	Panxian, Guizhou (102)	72.58	1932	Binchuan, Yunnan (102)	65.92
1932	Dingfan, Guizhou (100)	80.02	1932	Chuxiong, Yunnan (106)	78.15
1932	Zunyi, Guizhou (103)	59.49	1932	Yuanjiang, Yunnan (100)	82.96
1932	Dushan, Guizhou (100)	65.15	1932	Yuxi, Yunnan (104)	71.90
1932	Yiliang, Yunnan (100)	34.03	1932	Yongren, Yunnan (102)	45.47
Yangtze Rice–Wheat Area [70.64%]					
1930	Kunshan, Jiangsu (83)	65.97	1932	Zaoyang, Hubei (100)	83.81
1931	Yuyiao, Zhejiang (118)	32.35	1932	Yunmeng, Hubei (120)	73.70
1931	Hefei, Anhui (100)	63.93	1933	Huaiyang, Jiangsu (102)	86.12
1931	Yingcheng, Hubei (100)	79.65	1933	Fengyang, Anhui (100)	85.15
1932	Wuxi 2, Jiangsu (112)	87.21	1933	Wuhu, Anhui (100)	52.77
1932	Liu'an, Anhui (101)	65.17	1933	Pengze, Jiangxi (101)	71.81

Source: Calculations by the authors based on Buck's *China Land Utilization Statistics Information* and rural survey data

Note: Figures in parentheses are the quantity of farms, and figures in square brackets are agricultural labor surplus rates for the farming areas

an income as other industries could not absorb surplus labor. Due to the involution of the rural economy from its large population and limited land base, living conditions of farm households in China balanced on the edge of subsistence.[11] Peasants tried to input labor to their limited area of cropland and turn it into as much income as possible, so they could meet both production demands and the subsistence needs of their families. Therefore, farm households with more land resources faced less pressure to meet subsistence demands, inputted less labor, and had more free time. In contrast,

[11] Huang Zongzhi, 2000. *Small Peasant Economy and Social Change in North China* [M], Beijing: Zhong Hua Book Company.

Table 7.3 Per household cropland area and proportion of land rented in

Wheat zone			*Rice zone*		
Farming area	*Per household cropland area (mou)*	*Proportion of land rented in (%)*	*Farming area*	*Per household cropland area (mou)*	*Proportion of land rented in (%)*
Spring Wheat	38.04	12.16	Rice–Tea	13.64	48.56
Winter Wheat–Millet	19.76	19.63	Southern Rice	13.39	31.81
Winter Wheat–Kaoliang	27.30	18.50	Yangtze Rice–Wheat	16.23	54.93
Mean	28.37	16.76	Mean	14.42	45.10

Source: Calculations by the authors based on Buck's rural survey data

Note: Proportion of rented-in land = area of rented-in land/total area of cropland × 100%. Per household cropland area = total cropland area/total number of households

households with less cropland faced more pressures to meet subsistence needs, would increase the input of labor even under a very low marginal return of labor input, and had less free time. Moreover, the proportion of land rented in is another form of subsistence pressure faced by farm households. In the Republican era, the degree of land concentration was quite high. Landlords and rich peasants comprised less than 10% of the total population yet owned 70%–80% of rural land.[12] In Table 7.3 we summarize land rent. In a neoclassical setting with excess household/farm labor the shadow price of labor would be zero unless put to some alternative use in the wage market. With limited alternatives, renting land would not only increase agricultural output for the farm household, but would also increase labor demand, which in turn would reduce surplus. Providing rents were reasonable, the addition of rented lands to the farm base to a point at which labor surplus was removed would move the shadow price of labor from zero to some value greater than zero. However small that shadow price would be, for surely it was below the prevailing wage rate, the renting of land could be viewed as a strategy to reduce the labor surplus of some households. At the more aggregated

[12] *Policies and Regulations of the Ministry of Agriculture of PRC. Rural China over 40 Years* [M]. Zhengzhou: Zhongyuan Peasants Press, 1989. p. 107.

level, however, this would likely impact labor surplus. In the short run, the labor-to-land ratio would be fairly static. In this situation labor supply would be a zero sum game in the sense that the reduction in labor surplus of households renting land in would come at the expense of labor demand in that land's previous use.

As shown in Table 7.3, the per household cropland area in the wheat zone was larger than that in the rice zone, while the proportion of rented-in land in the rice zone was greater than that in the wheat zone. With larger areas of cropland, households in the wheat zone faced less pressure to meet subsistence demands, labor input into agricultural production was less, and thus the rate of surplus labor was higher. In contrast, in the rice zone, farm households had less cropland and faced higher levels of pressure to meet subsistence needs, a situation which was further worsened by rented-in land being more expensive: in percentage as high as about 45%. In such cases, as long as the input of labor could increase output or income, farm households would continue to input more labor into the land and give up leisure. Therefore, surplus agricultural labor in the rice zone was less than that in the wheat zone.

Cropping structures directly influenced the demand for agricultural labor in different regions. The demand for agricultural labor equaled the labor demand per unit crop area times total crop area. Buck's *Statistics Information* provides detailed statistical records on the labor demand of the unit crop area of different regions. Different crops required varying amounts of labor; in particular, cash crops required more labor than cereal crops. For example, one hectare of cotton and one hectare of peanuts required 248 days and 212.7 days, respectively, while for rice and wheat, one hectare required 185.4 days and 95.4 days, respectively. Regional differences in cash crops percentage directly influenced the demand of labor. In regions with a higher percentage of cash crops, there was a higher demand for labor and thus a lower rate of surplus agricultural labor. In contrast, regions with a lower percentage of cash crops had less demand for labor and thus the rate of surplus agricultural labor was higher.

In addition, the multiple cropping index (MCI) was another essential factor that influenced the total demand of labor (Table 7.4). The higher the MCI, that is, the greater area of sown land of crops, the greater the total demand for labor and thus the lower the rate of surplus agricultural labor. Conversely, the lower MCI, the higher the rate of surplus labor.

As presented in Table 7.4, both cash crops percentage and MCI were higher in the rice zone than in the wheat zone. Therefore, more laborers

Table 7.4 Percentage of crop area and MCI

Wheat zone			*Rice zone*		
Farming area	*Cash crops percentage (%)*	*MCI*	*Farming area*	*Cash crops percentage (%)*	*MCI*
Spring Wheat	12.15	0.93	Rice–Tea	18.67	1.51
Winter Wheat–Millet	12.25	1.12	Southern Rice	19.35	1.35
Winter Wheat–Kaoliang	19.62	1.21	Yangtze Rice–Wheat	26.28	1.57
Mean	14.67	1.09	Mean	21.43	1.48

Source: Calculations by the authors based on Buck's rural survey data

Note: MCI = total area of sown land for the crop/total cropland × 100%. Cash crops percentage = area of sown land for cash crops/total area of sown land × 100%

were required in the rice zone, resulting in a lower rate of surplus agricultural labor, as compared with the wheat zone. Moreover, the quantity of labor required per unit cropping area of rice, the main cereal crop in the rice zone, was nearly twice that of wheat. Therefore, crop variety was one of the major factors influencing the rate of surplus agricultural labor too.

7.5 Animal-Labor Substitution and Irrigation

Households' labor distribution in different regions was influenced by crop production patterns, including both labor-intensive and capital-intensive patterns (Table 7.5). In the Republican era, crop production was mostly labor intensive and labor animals substituted in part for manual labor. When farm households owned more labor animals, more labor was substituted out and this contributed to surplus labor. Likewise, with all other things being equal, households owning fewer labor animals would require more human labor and surplus labor would be less. In addition, different quantities of labor would be required for different production activities: irrigation required the most. The greater the percentage of irrigated area, the more labor time allocated by the household, resulting in less surplus labor.

The data presented in Table 7.5 shows that, per household, labor animal quantity was greater in the wheat zone than in the rice zone, with the exception of the Southwestern Rice Area. The substitution of a great quantity of labor animals for human labor was one of the main causes for a

Table 7.5 Per household labor animal quantity and percentage of irrigated area

Wheat zone			*Rice zone*		
Farming area	*Per household labor animal quantity*	*Percentage of irrigated area (%)*	*Farming area*	*Per household labor animal quantity*	*Percentage of irrigated area (%)*
Spring Wheat	1.78	40.25	Rice–Tea	0.67	80.83
Winter Wheat–Millet	0.74	8.92	Southern Rice	2.00	81.64
Winter Wheat–Kaoliang	0.89	18.02	Yangtze Rice–Wheat	0.65	55.64
Mean	1.14	21.34	Mean	1.11	72.70

Source: Calculations by the authors based on Buck's *Statistics Information* and rural survey data

Note: Percentage of irrigated area = irrigated crop area/total crop sown area × 100%. Per household labor animal quantity has been converted to standard labor animal unit

higher rate of surplus labor in the wheat zone than in the rice zone. The per household labor animal quantity in the Southwestern Rice Area was two, which was unusually high. However, as this area (Guizhou and Yunnan) is located in the Plateau Mountains, where farm fields are mostly terraced, animals were used mainly for transportation, with few being used in field production. The quantity of agricultural labor input was still quite great, therefore, and the rate of surplus labor was lower. Moreover, the percentage of irrigated area in the rice zone was far more than that in the wheat zone: "*In the rice zone, irrigated area exceeds two thirds of the total area, while in the wheat zone, irrigated area was small and scattered.*"[13] Therefore, in the rice zone households allocated more labor time to farm work, thus labor surplus was less.

7.6 Conclusions

Drawing on Buck's rural survey data, this chapter has used Buck's data to examine the surplus of agricultural labor in the rice and wheat zones of China, including six farming areas. Then, based on the estimation, this

[13] John L. Buck, 1936. *China Land Utilization* [M] Nanjing: Department of Agricultural Economy, Nanking University. p. 292.

chapter has discussed how some critical factors in agricultural production might have influenced labor surplus, including the endowment of agricultural production factors resources, cropping structures, and farming patterns. The analysis revealed that the rates of surplus agricultural labor in the rice zone and in the wheat zone were 69% and 79%, respectively. A large population with relatively little land used for agricultural production, less labor required per unit crop, a lower percentage of cash crops, lower MCI, less irrigated area, and more labor animals were the major factors influencing regional difference.

In conclusion, in the mid-Republican era there was a significant surplus of agricultural labor in rural areas of China. However, this surplus did not necessarily exist in the form of excess labor but, perhaps, more in the form of surplus time. Periodically, during harvest for example, laborers might have been fully employed. However, between major cropping activities those same laborers sat idle, contributing to surplus labor. This is consistent with the characteristics of agricultural production and differs from industrial production, which can be carried out continuously and evenly. Hence, labor surplus not only wasted labor, an important production factor, but also led to low economic efficiency for society as a whole and contributed to the poverty of peasants. The characteristics of traditional agricultural production and imperfection of labor markets impeded the transfer of surplus agricultural labor to non-agricultural employment to a great extent.

Acknowledgements This chapter was supported by a grant from National Social Sciences Foundation of China for the project Research on Construction of Agricultural Production System and Agricultural Products Consumption Pattern Characterized by Low Carbon Emission (Grant No. 10zd&031). It is also supported by a grant for the Construction of Buck's Survey Database, a Major Research Program of Humanities and Social Sciences, Nanjing Agricultural University (SKZD201201).

References

Buck, J.L. 1930. *Chinese Farm Economy: A Study of 2866 Farms in Seventeen Localities and Seven Provinces in China*. Chicago: University of Chicago Press.

Han, Jiafu. 1945. *Industrialization and Agricultural Construction of China*. Beijing: The Commercial Press.

Huang, Zongzhi. 2000. *Small Peasant Economy and Social Change in North China*. Beijing: Zhong Hua Book Company.

Jorgenson, W. 1967. Surplus Agricultural Labor and the Development of a Dual Economy. *Oxford Economic Papers* 19 (3): 288–312.

Lewis, W.A. 1989. *Theory of Dual Economy Beijing Economy*. Beijing: College Press.

Ministry of Agriculture of PRC. 1989. *Policies and Regulations of the Ministry of Agriculture of PRC. Rural China Over 40 Years*. Zhengzhou: Zhongyuan Peasants Press.

Schultz, Theodore W. 1999. *Transforming Traditional Agriculture*. Beijing: The Commercial Press.

Xue, Muqiao. 1946. *Elementary Knowledge on Rural Economy of China*. Dalian: Dazhong Bookstore.

Zhang, Li. 2007. Population, Land and Agricultural Productivity Level: Measurement of Population Pressure in Rural Wuxi in the Early 20th Century. *Agricultural History of China* 3: 47–46.

Zhong, Funing. 1995. Estimation of the Quantity of Surplus Labor in Rural Area of China. *Issues in Agricultural Economy* 9: 18–19.

Zhou, Zhongjian. 1997. Exploration on Supply and Demand of Agricultural Labor in Southern Jiangsu in Modern Times (1912–1937). *Agricultural History of China* 16 (3): 73–81.

CHAPTER 8

Agricultural Poverty and Inequality in 1930s China: Estimates of Gini and Engel Coefficients from Buck's Data

Hao Hu and Zhongwei Yang

8.1 Introduction

At present, increasing income inequality across different regions and industrial sectors of China have become a focus of both government and academic circles. A number of studies have been conducted on how economic reforms have affected income distribution, with a particular focus on regional inequality. There are two research dimensions for empirical studies on regional income inequality in China: one is the investigation of interprovincial income inequality by provinces, autonomous regions, and municipalities. The other is the investigation of income inequality across

H. Hu (✉)
College of Economics and Management, Nanjing Agricultural University, Nanjing, China
e-mail: huhao@njau.edu.cn

Z. Yang
Business School, Nanjing Xiaozhuang University, Nanjing, China

H. Hu et al. (eds.), *Chinese Agriculture in the 1930s*,
https://doi.org/10.1007/978-3-030-12688-9_8

the three major regions, that is, the East, Central, and West regions.[1] The former research argued for a U-shaped change with regard to regional income equalities, which decreased during 1978–1990 and increased after 1990. The latter research suggested that income inequality across the three major regions showed a trend of increasing. Several authors have conducted in-depth research on the structural decomposition of regional income inequality.[2]

While these studies have helped us better comprehend the law of income distribution and clarified the causes of income inequality in different regions and industrial sectors, they are rarely presented in a historical context. This is largely due to the unavailability of disaggregated data, particularly in pre-revolutionary China. Similarly, a lack of systematic micro-survey data on the income of agricultural hired labor and wages of rural residents before the foundation of the People's Republic of China (PRC) has greatly handicapped comparative studies on wages and income inequality of agricultural hired labor.

The era of rural reconstruction, which was launched in China during the 1920s–1930s, was the country's first attempt to transition its traditional—often backward—agricultural economy into a modern agricultural economy. With this movement, tremendous changes took place in all fields of rural economy, including productivity, economics of production, operational patterns, rural economic structure (including agricultural cooperative development and credit societies), and peasants' living conditions. A large number of people with scientific and economic insight, including John Lossing Buck and other faculty and students at the University of Nanjing, made significant strides in the development of rural areas in China.[3] Their practice and theories not only enabled people to better

[1] Wei Houkai 2002, Kanbur and Zhang 2005, Zhai Bin and Tong Haibin 2012.

[2] Lin et al. (1998), Lishi and Gustafsson (2008), Xue Jinjun (2008) and Gao Fan (2012).

[3] Among these efforts, *Chinese Farm Economy* and *Land Utilization in China* were the two most monumental works based on the large-scale rural economy survey organized by John L. Buck when he worked in the Department of Agricultural Economy, Nanking University. These two works "have not just rich materials but also fair and just arguments. All conclusions are completely based on survey data. So, their accuracy is unparalleled by any other common works that just express personal impressions" (Xie Jiasheng and Zhang Zhiwen, Preface for *Chinese Farm Economy*). Many scholars, both home and abroad, have highly complimented Buck's rural survey. Liang Fangzhong (1990) said: "Buck's survey is the first one in this field that tried to study such a broad and profound subject in such a comprehensive and systematic way." Since publication, these two works have continued to be regarded as classic works and have been widely cited by academic circles in Western countries, and in

comprehend society and national conditions at that time, but also provided valuable points of reference for rural reconstruction and agricultural modernization. Buck's survey period was from approximately the middle of 1929 through to the middle of 1933 and was thus at the cusp of the rural reconstruction period, so it is unclear to what extent these efforts had an impact on farmer respondents. Most certainly the full promise of reconstruction was far from fulfilled, but some trickledown effects in terms of improved seed, fertilizer, and credit may have had some influence in various localities, although we cannot make any further assertions beyond that.

Nonetheless, rural reconstruction was driven by the realization that China's agricultural economy was distressed in many ways—the persistence of poverty being one of them. From a historical perspective, and even for purposes of comparative economic analysis with present conditions in China's agricultural economy, it is important to understand how poverty and income inequality was dispersed in the mid-Republican era. To accomplish this, in this chapter we draw from Buck's data two measures of poverty and inequality, namely the Engel coefficient and the Gini coefficient.[4] The Engel coefficient measures the percentage of total household expenditures on food consumption. We measure this at the household level and make comparisons across China's three major regions. The Gini coefficient measures the dispersion of income across populations and areas, and is used to identify the range of income inequality. Buck does not record household income in *Land Utilization in China*, but does provide some measures of income in *Chinese Farm Economy*. With incomplete measures of household income we draw on Buck's measure of agri-

Taiwan and Hong Kong. For example, Huang Zongzhi (1986), Myers (1970), Arrigo (1986), and Zhang Wuchang (2000) obtained remarkable achievements in their research fields by utilizing Buck's survey to different extents. Nearly all studies on the rural society and economy of the Republic of China have regarded Buck's survey as providing the most important historical data. Now, in this new century, in response to the government's concern about issues on agriculture, countryside, and farmers, researchers have begun to review the development course of China's rural economy and expect to obtain enlightenment from its history. Buck and his survey on rural China are attracting increasing attention from researchers, with the survey results continuously reported (Shen Bangyao 2001; Li Jinzhen 2006; Ye Gongping 2007; Hu Hao and Zhong Funing 2008; Yang Xuexin and Ren Huilai 2010; Li Chunyan and Hu Hao 2012).

[4] Gastwirth 1972, Yao 1999.

cultural wage income as a proxy. We can justify this proxy in several ways. First, several chapters in this book have already discussed the labor problems in 1929–1933 China, including the issue of labor surplus. In a neoclassical sense, binding labor constraints would generate a shadow price on the next unit of labor to be hired. This shadow price measures the incremental gain in farm profits from all farming activities if one more unit of labor can be hired. In tight labor markets, this shadow price would capture the upper bound of the agricultural wage rate, in the sense that if agricultural wages exceeded the incremental gain in profits from additional labor, profits would actually increase. In a fluid labor market, with surplus labor, the supply of labor would exceed demand and so the agricultural wage might be substantially below its shadow price. In this way the economic translation of shadow price to wage rate will approximate agricultural conditions and incomes. If economic conditions strengthen, leading to better agricultural profitability, wage rates and agricultural wage incomes would increase, but if conditions were poor, with excess or surplus labor, the market clearing wage rate would be substantially lower. As for the Engel coefficients, we measure them directly from Buck's records on household expenditures.

In this chapter, we attempt to detail the income level of agricultural hired labor in different regions and regional inequality in 1930s China, as well as explore the causes of such inequality from the perspective of the income distribution trend of all China: East China, Central China, and West China. We use Gini coefficients and employ nonparametric Gaussian kernel density estimation methods using Buck's survey data.

8.2 Research Methodology and Data Processing

8.2.1 The Gini Coefficient

The Gini coefficient, as an internationally commonly used index for measuring statistical dispersion, is more commonly used to statically measure the overall equality or inequality of income disparity components under the influence of a variety of factors set over a certain period. It is usually applied to measure the distribution inequality of income, consumption, wealth, and other economic measures. The earliest estimation and application of the Gini coefficient with regard to an analysis of income inequality was that by Gastwirth (1972). Later, Yao (1999) presented the upper and lower limits of the Gini coefficient under grouped data conditions and continuously renewed the methodology of the measurement, making it more scientific and reasonable.

This chapter adopts the intuitive and simple income Gini coefficient calculation methodology proposed by Yao (1999) to estimate the coefficient of agricultural hired labor income across regions, with Eq. (8.1) as follows:

$$G = 1 - \sum_{i=1}^{n} W_i^* \left(2Q_i - P_i\right) \quad Q_i = \sum_{k=1}^{i} P_k \tag{8.1}$$

where P_i is the ratio of agricultural hired labor income of region i (i = 1, 2, ... n) to the total amount of agricultural hired labor income, W_i is the ratio of the population of region i to the total population, and Q_i is the accumulative total of P_i from k = 1 to i. Although this approach has the advantages of comprehensiveness and generality it also has certain limitations. The variation of the Gini coefficient is relevant with regard to the variation of income distribution across different regions, but it can neither describe the dynamic change of income distribution in spatial location, nor reflect the structural change of income inequality. The distribution dynamic method, put forward by Quah (1997), can make up for this shortcoming. This method takes the interregional income distribution pattern as a kind of probability distribution, describes the distribution shape using nonparametric kernel density estimation, and focuses on studying the dynamic trend of probability distribution in long-run development. In the course of distribution evolution, interregional economies will aggregate to different peaks, producing a "polarization" phenomenon. The distribution dynamic analysis method is data-driven. The most remarkable advantage of this method is that it does not need to assume the form of convergence or set a condition of convergence, yet it describes changes in regional disparity with a dynamic and intuitive distribution.

In this chapter we adopt a Gauss kernel density estimation method to reveal the distribution of agricultural hired labor income at the national level and in East China, Central China, and West China. In estimation, data needs to be normalized before analysis is conducted. The logarithm of the ratio of y_i to $\overline{y}$ is used as a variant x_i to normalize the data, where y_i is the per capital index value of agricultural hired labor income of region *i*, and $\overline{y}$ is the per capital index value of agricultural hired labor income at the national level. With this, x_i is expressed, as in Eq. (8.2):

$$x_i = \ln\left(\frac{y_i}{\overline{y}}\right) = \ln y_i - \ln \overline{y} \tag{8.2}$$

The Gauss kernel density function is expressed in Eq. (8.3):

$$\overline{f}(x) = \frac{1}{nh}\sum_{i=1}^{n}\frac{1}{\sqrt{2\pi}}e^{\left(-\frac{1}{2}\left(\frac{x-x_i}{h}\right)^2\right)} \tag{8.3}$$

where the observed value of $w(x_i)$ is obtained by calculation, n is the number of samples investigated, h is the bandwidth of a function used for adjusting the smoothness of distribution curves, and x is the range of the x-coordinates. For easy analysis, the value of h is fixed at 0.27, and the range of x is set at (−2, 2).[5] This Gauss kernel function ignores consideration of population weight, while in practice the regional disparity of agricultural hired labor income is ultimately embodied in people. As pointed out by Gisbert (2003), when nonparametric methods are used to study convergence, the results may be misleading if no consideration is given to the influence of regional population differences on convergence. Therefore, the Gauss kernel density function is modified on population weight.

The Gauss kernel density function in this chapter, after being modified by the ratio of the population of different provinces to the total population of China, is expressed as follows:

[5] To select the h value, the degree of curve smoothness as well as the necessity to compare and analyze the curves for the years selected will be taken into account. After balancing these two factors, this chapter found after calculation that the square error of fitting is minimum when 0.27 is selected as the h value. The calculation equitation is expressed as: $p(h) = n^{-1}\sum_{i=1}^{n}\left(m(x_i) - \hat{m}_n(x_i)\right)^2 w(x_i)$, where $\hat{m}_n(x_i)$ is the sample weighted average in a domain near x, and $w(x_i)$ is the weight. The range of X is determined by the interval of x_i value.

$$\bar{f}(x)=\frac{1}{h}\sum_{i=1}^{n}W_i\frac{1}{\sqrt{2\pi}}e^{\left(-\frac{1}{2}\left(\frac{x-x_i}{h}\right)^2\right)} \tag{8.4}$$

where W_i is the ratio of the population of different provinces to the total population of China, and $\sum_{i=1}^{n}W_i=1$.

8.2.2 Data Sources and Processing

Data used in this study are data on income of agricultural labor hired on a one-year basis from Buck's 1929–1933 survey, arranged by provinces, cities, and counties.[6] To further study the income disparity of agricultural hired labor of different regions of China, we adopt the regional division criteria as set forth in Document (2000)33 issued by the State Council of China, which divides China into three regions, namely, East China, Central China, and West China, and process Buck's survey data accordingly.[7]

[6] Specifically, such data includes income data for the following regions (109 cities and counties across 18 provinces): Hebei Province (Fuping, Changli, Zhengding, Jiaohe, Nangong, Xushui, Cangxian, Qingxian, and Tongxian), Shandong Province (Enxian, Fushan, Huimin, Yishui, Laiyang, Zibo, Ningyang, Tai'an, Tangyi, Jimo, Qingning, and Yixian), Jiangsu Province (Guanyuan, Changshu, Wujin, and Wuxi), Zhejiang Province (Yuyao, Fenghua, Huangyan, Linhai, Lishui, Chun'an, Tangxi, Tonglu (1) and Tonglu (2), Dongyang, and Yongjia), Guangdong Province (Zhongshan, Gaoyao, Maoming, and Nanxiong), Henan Province (Linbao, Luoyang, Jixian, Xiangcheng, Kaifeng, Linzhang, Shangqiu, Qinyang, Yancheng, and Xinyang), Shanxi Province (Anyi, Linxian, Pingding, Shouyang, Jincheng, Qingyuan, Wuxiang, Ningwu, and Jingle), Anhui Province (Hefei, Liu'an, Dingyuan, Tongcheng, Wuhu, and Xiuning), Hubei Province (Anlu, Hanchuan, Huangpu, Yingcheng, and Yunmeng), Hunan Province (Changsha, Changde, Yiyang, Chengxian, and Xinhua), Jiangxi Province (Nanchang, Pengze, Duchang, Fuliang, and Gao'an), Guangxi Province (Guilin, Liucheng, and Yongning), Shanxi Province (Zhen'an, Zhouzhi, Shangxian, Dingbian, and Yulin), Gansu Province (Pingliang, Tianshui, Gaolan (2), and Wuwei), Qinghai Province (Huangyuan and Xining), Sichuan Province (Chongzhou, Chongqin, Fengdu, Fuling, Xinfan, Luzhou, Mianyang, and Daxian), Guizhou Province (Anshun, Panxian, Dingfan, Zunyi, and Dushan), and Yunan Province (Binchuan and Yuanjiang).

[7] Division criteria: The East China region includes 11 provinces and municipalities directly under the central government, namely, Beijing, Tianjin, Hebei, Liaoning, Shanghai, Jiangsu, Zhejiang, Fujian, Shandong, Guangdong, and Hainan. The Central China region includes

8.3 China's Agricultural Hired Labor Income Level and Inequality in the 1930s: A Gini Coefficient Perspective

As shown in Table 8.1, the Gini coefficients for the income of agricultural hired labor in the different regions of China is estimated using Eq. (8.1). The average income of agricultural hired labor for different provinces are processed accordingly. In the course of processing, we considered the influence of the population weights of cities and counties in different provinces. Consequently, the average incomes of agricultural hired labor presented in Table 8.1 are weighted averages, with the populations of different counties and cities used as weights, instead of simple arithmetic averages.

Whether for a country or region, the level of income is always closely related to the level of economic development. Due to the availability of indexes and data, this chapter uses Engel coefficients for farm households to describe the economic development levels of some of the different regions of China in the 1920s and 1930s (see Table 8.2).

The Engel coefficient is the ratio of the amount of food expenditure to the total amount of expenditure and is commonly used to measure the influence of economic development and income increases on food consumption. It is one of the major indictors used to measure the richness of a family or a country. A higher Engel coefficient indicates lower income for a family or more backward economic development for a country. Conversely, the lower the Engel coefficient, the more developed the economy or the higher the resident income relative to food expenditures. As we can see from the results in Table 8.2, the Engel coefficients for East China are lower than those of Central China, indicating that during the mid-Republican era the economic development level in the former was higher than that in the latter.

Combining Engel coefficients with the income level and Gini coefficients of agricultural hired labor of these regions, we find that the levels of average agricultural hired labor income for East China, Central China, and

eight provinces, namely, Shanxi, Jilin, Heilongjiang, Anhui, Jiangxi, Henan, Hubei, and Hunan. The West China region includes 12 provinces, autonomous regions, and municipalities directly under the central government, namely, Chongqing, Sichuan, Guizhou, Yunnan, Tibet, Shaanxi, Gansu, Qinghai, Ningxia, Xinjiang, Guangxi, and Inner Mongolia.

Table 8.1 Average agricultural hired labor income levels and Gini coefficients for different provinces and regions (income units in silver dollars)

Region	*Average income*	*Gini coefficient*	*Region*	*Average income*	*Gini coefficient*	*Region*	*Average income*	*Gini coefficient*
East	103.5188	0.2242	Central	84.7025	0.2103	West	65.6905	0.2177
Hebei	87.6324	0.1049	Henan	50.7684	0.1159	Guangxi	106.2480	0.2137
Shandong	83.9513	0.1302	Shanxi	86.6929	0.1908	Shaanxi	50.7466	0.1812
Jiangsu	104.7296	0.1225	Anhui	81.1081	0.1292	Gansu	59.9069	0.2802
Zhejiang	105.0513	0.1276	Hubei	113.1321	0.1498	Qinghai	46.5824	0.0109
Guangdong	154.1603	0.3388	Hunan	101.6297	0.1493	Sichuan	57.5804	0.1190
–	–	–	Jiangxi	101.0198	0.0926	Guizhou	76.7511	0.1138
All China	89.4396	0.2360	–	–	–	Yunnan	144.0322	0.0760

Note: Silver dollar is the main active money and unit of price under the silver standard monetary system in1930s

Table 8.2 Engel coefficients for farm households in different regions of 1920s–1930s China

Survey locality	*Year*	*Number of households*	*Average annual expenditure per household (Yuan)*	*Average annual food expenditure per household (Yuan)*	*Engel coefficient*
East China					
Pingxiang, Hebei	1923	152	88.62	58.83	66.38
Yanshan, Hebei	1922	150	113.13	62.20	54.98
Yanshan, Hebei	1923	133	155.20	88.02	56.71
Chunhua, Jiangning, Jiangsu	1923	203	338.80	179.56	53.00
Taipingmeng, Jiangning, Jiangsu	1923	217	251.33	123.55	49.16
Tushan, Jiangning, Jiangsu	1934	286	228.15	136.36	59.77
Wujin, Jiangsu	1924	300	293.26	191.99	65.47
Lianjiang, Fujian	1922	161	336.69	178.27	52.95
Total or average of East China		1602	225.65	127.35	56.44
Central China					
Xinzhen, Henan	1923	144	258.65	194.31	75.12
Kaifeng, Henan	1923	149	349.67	268.16	76.69
Wuxiang, Shanxi	1922	251	115.34	57.64	49.97
Huaiyuan, Anhui	1924	124	185.16	107.17	57.88
Suxian, Anhui	1923	286	259.26	153.48	59.20
Lai'an, Anhui	1922	100	223.06	108.58	48.68
Total or average of West China		1054	231.86	148.22	63.93
National total or average		2656	228.31	136.29	59.70

Source: Authors' calculations based on Yan Xinzhe (1993), *Rural Households Survey*, pp. 110–115 and Buck (1936), *Chinese Farm Economy*, p. 528. The title of this table has been changed slightly by the authors as the term "Engel coefficient" was not used in the original. The average annual expenditure per household includes expenditures for food, clothing, housing, fuel, and miscellaneous expenses

West China showed a gradual decreasing trend consistent with regional economic development levels. Specifically, the income of agricultural hired labor was higher in East China (approximately twice those in West China) where the economic development level was higher; the level of economic development was less high in Central China and was lowest in West China, where development was relatively backward. For example, in provinces within East China such as Guangdong, Zhejiang, and Jiangsu, the income of agricultural hired labor was higher than in provinces such as Qinghai, Shaanxi, and Sichuan in West China. Yunnan Province and Guangxi Province, though located in West China, had a higher level of agricultural hired labor income. Moreover, different cities and counties within Yunnan Province saw a very small difference in income of agricultural labor hired by year.[8] From Table 8.1 we also found that Gini coefficients are great for both eastern provinces such as Guangdong and Zhejiang, where the income level of agricultural hired labor was high, and western provinces such as Gansu and Shaanxi, where the income level of agricultural hired labor was low, suggesting greater inequality between different cities and counties within the provinces.

Observing the Gini coefficients, the regional economic development level and intraregional income inequality of agricultural hired labor roughly followed a positive U-shaped trend. The income inequality of agricultural hired

[8]The higher level of income for the agricultural hired labor of Yunnan Province in the 1920s–1930s was attributed partly to the improvement of production forces. As soon as the Republic of China had been established, the local government adopted a policy to develop the economy by vitalizing crafts, reorganizing industry, and supporting commerce. After Long Yun seized the reins of Yunnan in 1929, peace and stability in the internal and external environment was restored to a certain extent, and the local government focused again on economic construction and built many plants. As recorded in *An Overview of Yunnan*: "In the recent 7–8 years, politics has gradually returned to the right track, social order has become steady, finance order has been rectified and the number of plants are increasing" (Yunnan Branch Office of Nanking-Yunnan Road Tour Survey Preparation Office. (1937). *An Overview of Yunnan*, p. 65). On the other hand, "Objectively, the opening of Yunnan–Vietnam Railway in the beginning of the twentieth century provided a historical turning point for social and economic development of Yunnan Province" (Chen Zhengping (2002, p. 98). The relatively smaller inequality in the income of the agricultural hired labor of Yunnan may be explained by the fact that: (1) the government policy did not cause great inequality across different cities and counties within the region, (2) the survey data did not cover all parts of Yunnan Province and thus did not reflect the overall situation of agricultural hired labor income in the province. As for Guangxi Province, the higher level of agricultural hired labor income was mainly attributed to a series of policies implemented in the 1920s–1930s by the New Guangxi Clique headed by Li Zongren and Bai Chongxi. These policies included implementing the "Trinity" system and villagers' autonomy, establishing rural cooperatives, developing the rural economy, promoting compulsory education and adult education to improve citizens' intelligence and knowledge levels, advocating new ways of living, and changing old customs, and developing facilities for medical services, public health, agriculture, forestry, and water conservation to improve production and living conditions. A series of rural reform and construction initiatives led to a new look for the countryside of Guangxi, in contrast to other provinces, and greatly promoted the social development of this area.

labor was at a higher level in economically better-developed East China, lower in West China with backward economic development, and lowest in Central China. In East China, the resident income level and consumption level were, accordingly, higher. However, there was also inequality in economic development within the region, leading to worse inequality of agricultural hired labor income, compared to that in Central China and West China. The income inequality in West China was great too. This could be related to backward economic development in West China, where insufficient internal and external demands severely impeded economic growth, leading to an overall backward economy with low agricultural wages. However, in a few provinces in the area the income of agricultural hired labor greatly exceeded the overall average income of West China as a whole, leading to significant intraregional inequality in agricultural wages within West China. In the Central China region, though the economic development level was not the highest, interregional economic development was relatively even and the income distribution was relatively reasonable, when compared with China generally and East China and West China more specifically. We also observed that, despite significant interregional disparity in economic development in 1920s–1930s China, regional inequality with regard to the income of agricultural hired labor did exist but polarization did not occur.[9]

Gini coefficients intuitively provide us with an understanding of the degree of inequality in income of agricultural hired labor, but what about agricultural hired labor income distribution across different cities and counties in East China, Central China, and West China over the 1930s? Obviously, this question cannot be answered using Gini coefficients alone. Further study on this question is conducted using the Gauss kernel density analysis method.

[9] This means that there was no polarization trend in the income of agricultural hired labor across different regions. The main cause of this is that, in the vast rural areas of 1920s–1930s China, there was little change in the level of social production forces. There was no remarkable development of agriculture in rural areas, where ancient tools of production such as hoes, plows, harrows, and water wheels were still being widely used; inefficient traditional production methods were still being applied; production forces developed very slowly; and the self-sufficient natural economy of household crafting and peasant farming occupied a key position in the overall rural economy. Consequently, there was absolutely no condition for polarization of income of agricultural hired labor at the bottom of society.

8.4 Agricultural Hired Labor Income Distribution over the 1930s: A Perspective of Kernel Density Distribution

Using Eq. (8.4) we obtained the distribution curves of agricultural hired labor income for all China, East China, Central China, and West China (as shown in Fig. 8.1). By analyzing the kernel density distribution curves of such income for these areas, we may reach the following conclusions:

First, the income distribution of agricultural hired labor across different provinces of China in the 1930s roughly presented a unimodal distribution trend. According to Quah (1997), unimodal distributions suggest that the income of agricultural hired labor across different regions of China had formed a uniform convergence club. Therefore, at the national level there was no remarkable disparity in such income across different provinces. In addition, on the right side of the kernel density distribution curve of agricultural hired labor income for all China there is an unremarkable small peak, showing that this income had formed a uniform convergence club across provinces with relatively higher income levels.

Second, as observed from the shape of the density curves, there is significant disparity between the distribution curves of agricultural hired labor income for East China, Central China, and West China. The distribution curve for West China, similar to that for all China, presents a unimodal distribution trend. This indicates that the income of agricultural hired labor in different provinces of Central China had formed a uniform convergence club without remarkable disparity. Although, for both East China and West China, their kernel density distribution curves present characteristics of multi-club convergence, their patterns of multi-peak distribution are opposing. As shown in Fig. 8.1, the distribution curves for both East China and West China form a small peak in the lower income section and the higher income sections, respectively. Though on a scale far smaller than the main peak, the small peaks steadily exist over the investigated period, implying the existence of both low-income and high-income clubs. Specifically, for East China, remarkable peaks are formed in both the medium-income section and higher income section. In contrast, for West China, the peaks formed in the lower income section and medium-income sections are remarkable, indicating that there were both very high-income provinces and very low-income provinces within East China and West China, that is, there were a few provinces with very high incomes in East China and a few with very low incomes in

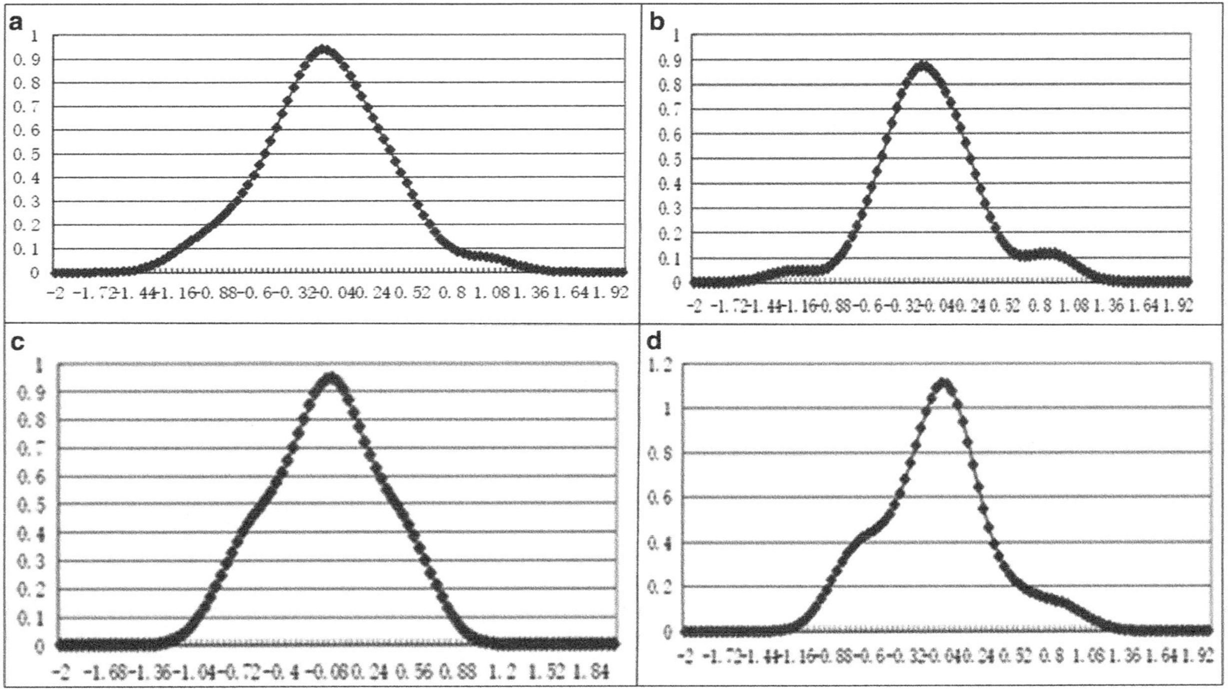

Fig. 8.1 Income distribution of agricultural hired labor in all China, East China, Central China, and West China. (**a**) Income distribution of agricultural hired labor in all China. (**b**) Income distribution of agricultural hired labor in East China. (**c**) Income distribution of agricultural hired labor in Central China. (**d**) Income distribution of agricultural hired labor in West China. Note: The horizontal axis represents normalized income of agricultural hired labor, the vertical axis represents probability density value

West China. This is an important contributor for the great disparity in agricultural hired labor income within East China and West China.

Finally, the inequality of agricultural hired labor income at the national level can generally be attributed to disparities between East China, Central China, and West China. As shown, the Gini coefficients of agricultural hired labor income for East China, Central China, and West China are all lower than that for the whole country. This is mainly due to the coexistence of laborers with a very low income and laborers with a very high income in the labor markets of East China, Central China, and West China. However, this coexistence was not likely to expand disparity in such incomes within a region due to the spatial proximity effect. But, if taken as a whole, it remarkably expanded the interregional disparity in agricultural hired labor income.

Obviously this kernel density analysis provided more abundant information than the Gini coefficient method. It can describe more vividly the distribution of disparity in the income of agricultural hired labor across different regions of China. The multi-peak distribution of kernel density analysis particularly, which shows convergence within developed regions and undeveloped regions, as well the expansion of inequality in the income of agricultural hired labor across different regions, can satisfactorily settle disputes in previous studies on the nature of convergence.

8.5 Summary

In this chapter, we quantitatively calculated the average income of agricultural hired labor and the income inequality across different regions of China by introducing Gini coefficient methodology and using Buck's survey data for the 1930s. We also analyzed the distribution and influence mechanisms of agricultural hired labor income using the nonparametric Gauss kernel density estimation method. Our study revealed the following findings: First, in the 1930s, there was remarkable inequality in economic development across different provinces, cities, and counties within East China, Central China, and West China. Very low-income labor and very high-income labor coexisted in the labor markets, although regional disparity did exist, albeit without polarization. Second, the average income of agricultural hired labor in East China, Central China, and West China showed a gradually decreasing trend, while the intraregional inequality and the regional economic development level presented a positive U-shaped trend. Specifically, the inequality in agricultural hired labor income was highest in the economically better-developed East China, lower in West China, with its backward economic development, and lowest in Central China. Third, the inequality in economic development across the three areas was an important factor for great disparity in the income of agricultural hired labor.

This study has two main defects: First, samples used in this chapter, though including data on the income of agricultural hired labor for as many as 109 cities and counties in 18 provinces, do not cover all the provinces of China, hence could not comprehensively present the income disparity for agricultural hired labor across all regions of China in the 1930s. Second, due to data unavailability, this study is only a static analysis based on cross-sectional data; the dynamic analysis method is not used to present the income distribution and inequality evolution trends of agricultural hired labor across different regions of China in the 1930s. Unfortunately, given the nature of the data available to us, we find no remedies for these.

References

Arrigo, L.G. 1986. Landownership Concentration in China: The Buck Survey Revisited. *Modern China* 12 (3): 259–360.

Buck, J.L. 1936. *Chinese Farm Economy.* Beijing: The Commercial Press.

Chen, Zhengping. 2002. *Study on the Course of Early Industrialization of Yunnan.* Beijing: Nationalities Press.

Gao, Fan. 2012. Decomposition of China's Resident Income Disparities Variation Factors: Trend and Explanation. *Economic Sciences* 3: 5–17.

Gastwirth, J.L. 1972. The Estimation of the Lorenz Curve and Gini Index. *Review of Economics and Statistics* 54 (3): 306–316.

Gisbert, F.J.G. 2003. Weighted Samples, Kernel Density Estimators and Convergence. *Empirical Economics* 28 (2): 335–351.

Hu, Hao, and Funing Zhong. 2008. Research on the Change in China's Agricultural Management over Nearly One Century. *Problems of Agricultural Economy* 110 (7): 15–22.

Huang, Zongzhi. 1986. *Peasant Economy and Social Change of North China.* Shanghai: Zhonghua Book Company.

Kanbur, R., and X. Zhang. 2005. Fifty Years of Regional Inequality in China: A Journey Through Central Planning, Reform and Openness. *Review of Development Economics* 9 (1): 87–106.

Li, Jinzhen. 2006. Rural Finance in Chinese History. *Capital Market* 2: 87–88.

Li, Chunyan, and Hao Hu. 2012. Change in Farm Households Social Intercourse Consumption Behavior Over the Past Century—Comparison Based on Buck's Survey Data. *Agricultural Archaeology* 12: 106–110.

Liang, Fangzhong. 1990. *Collection of Papers on History of Economy by Liang Fangzhong*. Guangzhou: Guangdong People's Press.
Lin, Yifu, Fang Cai, and Zhou Li. 1998. Analysis on Regional Disparity of China's Economic Transition. *Economic Research* 10: 3–10.
Lishi, Staly, and Bjön Gustafsson. 2008. *Research on China's Resident Income Distribution (III)*. Beijing: Normal University Publishing Press.
Myers, R.H. 1970. *The Chinese Peasant Economy: Agricultural Development in Hopei and Shantung, 1890–1949*. Cambridge, MA: Harvard University Press.
Quah, D.T. 1997. Empirics for Growth and Distribution: Stratification, Polarization, and Convergence Clubs. *Journal of Economic Growth* 2 (1): 27–59.
Shen, Bangyao. 2001. New Comprehension of Buck's China Rural Survey. *Academia Bimestris* 2: 119–123.
Wei, Houkai. 2002. Influence of Foreign Direct Investment on the Growth of China's Regional Economy. *Economic Research* 4: 19–26–92–93.
Xue, Jinjun. 2008. *Inequality of China: Research on Income Distribution Gap*. Beijing: Social Science Academic Press.
Yan, Xinzhe. 1993. *Rural Households Survey*. Beijing: The Commercial Press.
Yang, Xuexin, and Huilai Ren. 2010. Research on Change of Chinese Farm Tools Over 80 Years—Based on Buck's Rural Survey on 150 Households of Yanshan County in 1923. *Agricultural Archaeology* 8: 169–177.
Yao, S.J. 1999. On the Decomposition of Gini Coefficients by Population Class and Income Source: A Spreadsheet Approach and Application. *Applied Economics* 31 (10): 1249–1264.
Ye, Gongping. 2007. Buck's China Rural Survey. *Read* 12: 41–45.
Zhai, Bin, and Haibin Tong. 2012. Empirical Study on Peasant Income Inequality in the East, Central and West China Regions—Analysis from the Perspective of Income Sources. *Inquiry into Economic Issues* 8: 7–12.
Zhang, Wuchang. 2000. *Tenant Theory- Agriculture Applied to Asia and Land Reform in Taiwan*. Beijing: Commercial Press.

CHAPTER 9

An Analysis on the Inverse Relationship between Yield and Farm Size in Rural China in the 1930s

Hisatoshi Hoken and Qun Su

9.1 Introduction

It is widely believed that the distributions of landholdings were rather unequal and increasingly polarized during the late Qing dynasty and the Republican period. Tenancy farming appeared to give rise not only to social economic inequality among farmers, but also to have a significant impact on the efficiency of agricultural production. However, Myers (1976) utilizes detailed survey data conducted by the Provincial Industrial Investigating Bureau of Manchuria to investigate the socioeconomic structure in rural villages. Contrary to prevalent understandings of the late Qing dynasty and the Republican periods, he observed that there had been neither a trend toward social class polarization nor one toward more

H. Hoken (✉)
Kwansei Gakuin University, Nishinomiya, Japan
e-mail: suqunqun@njau.edu.cn

Q. Su
College of Economics and Management, Nanjing Agricultural University, Nanjing, China

H. Hu et al. (eds.), *Chinese Agriculture in the 1930s*,
https://doi.org/10.1007/978-3-030-12688-9_9

unequal land ownership or land use in Manchuria. Yanagisawa (2000) confirms that land polarization and land dispersion had occurred; consequently, land concentration by landowners was not particularly strong in that era of northern China. Rawski (1989) also indicates that the commercialization and specialization of farming had been gradually diffused in that period with the development of domestic industrialization and international export. These studies suggest the importance of reconsidering the prevalent images of prewar China.

Understanding the socioeconomic structure of rural farmers in China during the late Qing dynasty and the Republican period is important, and until now statistical examinations have been insufficient. In this chapter we use Buck's data to estimate cropping patterns and landholdings gaps in rural China in the 1930s. In addition, we also tackle the classical hypothesis as to whether there exists an inverse relationship between land productivity and cultivated area, as has been observed in other developing countries.

9.2 Cropping Patterns in the Republican Era

In this section we extract and summarize from the available microdata the crops grown across 24 counties in ten provinces; however, only 20 counties were useable. In Table 9.1 we identify the numbers of different crops grown by county. Entries of "N/A" represent counties for which data were recovered but were not complete enough to be used in this chapter. The total number of crops cultivated in 20 counties amounted to 377, including duplicate crops, suggesting a diversified agricultural economy during the 1930s. More than ten crops were cultivated in most counties. In Yuyao, Zhejiang, 39 varieties of crops were cultivated. Table 9.2 looks at this data in a different way by breaking down the crops and showing how many counties they were grown in. On the one hand, main crops such as wheat and rice were grown in many counties. Most of the non-grain crops were cultivated in only limited localities and grown by few farmers. For the purpose of simplifying our task, we selected crops which were grown by a relatively large number of farmers and constructed a micro database of these. They are identified in column 4 of Table 9.1, covering 45.6 percent of total varieties.

Table 9.1 Number of crops grown by county and selected crops for analyses

Province	*County*	*Number of crop items*	*Number of selected crop items*	*%*
Anhui	Fengyang	16	11	68.8%
	Fuyang	19	12	63.2%
	Ho	24	8	33.3%
	Wuhu	11	5	45.5%
Zhejiang	Tangki	23	15	65.2%
	Yuyao	39	11	28.2%
Henan	Nanyung	20	11	55%
	Sinyang	N/A	N/A	N/A
Hebei	Kishui	20	8	40%
	Yincheng	14	6	42.9%
	Yunmeng	N/A	N/A	N/A
Gansu	Ninghua	22	11	50%
	Wuwei	17	9	52.9%
Jianxi	Nanchang	6	1	16.7%
	Pengtse	23	7	30.4%
	Tuchang	26	15	57.75%
Jiangsu	Kwanyun	19	8	42.1%
	Kunshan	18	7	38.9%
	Wusih	6	3	50%
	Yencheng	5	4	80%
Guangdong	Chungshan	N/A	N/A	N/A
Shanxi	Ningwu	21	8	38.1%
Sichuan	Peikiang	28	12	42.9%
	Ta	N/A	N/A	N/A
Total		377	172	45.6%

9.3 Examination of Inverse Relationship between Land Productivity and Cultivated Area

In this section, we discuss cropping patterns and farm management in the Republican era. In particular, we seek to understand the relationship between farm size and crop yield and to statistically test the classical hypothesis on whether an inverse relationship exists between land productivity and cultivated area, as observed in other developing countries.

9.3.1 Characteristics of Cropping Patterns of Surveyed Counties

Since the conditions of agricultural production and the variation of crops differ considerably among counties, it is necessary to discuss in detail the characteristics of cropping patterns on surveyed areas. In order to examine

Table 9.2 Crops grown in selected counties

Item	*Number of counties*
Wheat	16
Barley	15
Rice	11
Cotton seeds	8
Rapeseed	8
Kaoliang	6
Millet	6
Broad bean	6
Green beans	6
Soybeans	6
Field peas	6
Sweet potatoes	4
Peanut	4
Early glutinous rice	3
Sesame	3
Winter radish	3
Black soybean	3
Barley, hulless	2
Spring wheat	2
Late rice	2
Millet	2
Cotton	2
Cotton lint	2
Opium	2
Barley field peas	2
Corn	1
Buckwheat	1
Glutinous rice	1
Early rice	1
Late glutinous rice	1
Hemp	1
Bauica	1
Jelacca	1
Green beans	1
Sesame	1
Buckwheat	1
White lady beans	1
Thyme	1
Lady beans	1

(*continued*)

Table 9.2 (continued)

Item	*Number of counties*
Wild lady beans	1
Sesame	1
Seed bottom	1
Sugar cane	1
Garlic	1
Eggplant	1
Mulberry	2
Astragalus	1
Red beans	1
Watermelon	1
Lentil	1
Brassica pekinensis	1
Millet+black bean	1
Potatoes	1
Carrots	1
Mullein leaves	1
Oats	1
Field beans	1
Hyacinth beans	1
Rapeseed	1
Unidentified	6

the variations of crops and the cropping patterns, we calculate the average of total cultivated areas and the percentages of cultivated areas of each crop to total cultivated area. In so doing, we classify crops cultivated in surveyed counties into six categories: grain, legume, oil crop, cotton, root vegetable, and others. The results are shown in Table 9.3.

The averages of total cultivated land differed strikingly among counties. The largest of all counties was Kishui in Hebei province, where the average amounted to 73 mou per household, while the smallest was 8.8 mou in Wusih, Jiangsu.

The land shares of each group to total cultivated areas are reported in Table 9.4. Grain accounts for more than 50 percent in every county except for Yuyao, Zhejiang, while the proportions of other types of crops, that is, legumes, oil crops, and cotton, were relatively high. Therefore, the

Table 9.3 Percentage share of cultivated areas of each crop type to total cultivated area

County	*Total cultivated areas (mou)*	*Grain (%)*	*Legume (%)*	*Oil crop (%)*	*Cotton (%)*	*Root vegetables (%)*	*Other (%)*
Fengyang	41.1	60%	15%	3%	0%	0%	22%
Fuyang	37.2	61%	28%	0%	2%	7%	2%
Ho	22.3	75%	3%	0%	19%	2%	0%
Wuhu	29.3	70%	0%	30%	0%	0%	0%
Tangki	17.6	56%	0%	3%	0%	0%	41%
Yuyao	29.4	1%	34%	3%	46%	0%	15%
Nanyung	40.7	90%	5%	0%	2%	0%	2%
Kishui	73.0	85%	0%	4%	10%	0%	0%
Yincheng	44.3	94%	0%	4%	3%	0%	0%
Ninghua	19.2	80%	6%	0%	0%	0%	14%
Wuwei	19.3	75%	3%	0%	0%	0%	22%
Nanchang	20.3	100%	0%	0%	0%	0%	0%
Pengtse	12.0	51%	4%	22%	23%	0%	0%
Tuchang	18.2	60%	13%	22%	1%	3%	2%
Kwanyun	66.2	66%	32%	0%	9%	1%	0%
Kunshan	37.0	82%	7%	4%	8%	0%	0%
Wusih	8.8	85%	0%	0%	0%	0%	15%
Yencheng	25.3	100%	0%	0%	0%	0%	0%
Ningwu	33.7	63%	7%	4%	9%	0%	25%
Peikiang	32.4	54%	18%	0%	2%	21%	5%

productions of commercial crops had already developed at this time, an observation that is consistent with the studies of Cao (1996) and Rawski (1989), while the cultivation of grains was dominant in rural China in the 1930s.

To elucidate the diversifications of crop productions, we calculate the average number of crops cultivated by each household. A crop raised by a farmer is counted as "one" item regardless of the size of cultivated area, and the averages of the total items by county is calculated.[1] Figure 9.1 shows that the average number of crop items cultivated by each farmer is over three in most counties, and most farmers concentrated on four to five items. In addition, we can see that the diversification of cropping was highly advanced in some localities, that is, over six varieties of crops were raised in four counties.

[1] When we count the varieties of crop cultivated, we regard "other crops" as one crop. Although this might induce underestimations of the crop variation, the share of farmers who cultivated other crops is not particularly high, so this method is acceptable.

Table 9.4 Percentage share of cultivated areas of grain crops

County	*Wheat (%)*	*Rice (%)*	*Barley (%)*	*Kaoliang (%)*	*Millet (%)*	*Other grains (%)*
Fengyang	67%	0%	8%	25%	0%	0%
Fuyang	65%	0%	1%	32%	1%	1%
Ho	21%	73%	6%	0%	0%	0%
Wuhu	6%	92%	2%	0%	0%	0%
Tangki	15%	28%	33%	9%	20%	4%
Yuyao	0%	0%	96%	0%	4%	0%
Nanyung	36%	56%	8%	1%	0%	0%
Kishui	12%	88%	0%	0%	0%	0%
Yincheng	5%	62%	32%	0%	0%	0%
Ninghua	49%	0%	0%	3%	48%	0%
Wuwei	50%	0%	12%	0%	38%	0%
Nanchang	0%	100%	0%	0%	0%	0%
Pengtse	14%	61%	25%	0%	0%	0%
Tuchang	7%	60%	18%	0%	15%	0%
Kwanyun	53%	0%	14%	14%	0%	20%
Kunshan	30%	61%	9%	0%	0%	0%
Wusih	39%	61%	9%	0%	0%	0%
Yencheng	5%	84%	11%	0%	0%	0%
Ningwu	5%	0%	0%	0%	0%	95%
Peikiang	13%	67%	11%	9%	0%	0%

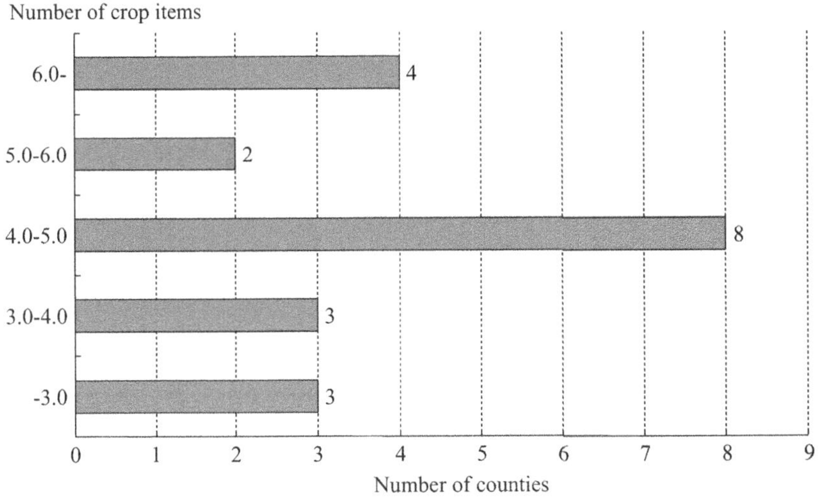

Fig. 9.1 Average crop items cultivated by farmers

As described in Table 9.3, however, the percentage share of cultivated areas of staple crops such as grain to total cultivated areas is dominant, thus we also estimate the county average of percentage share of the top three crops to total cultivated areas by farmers. Figure 9.2 clearly shows that the share of the top three crops surpasses 60 percent in every county, and most of the percentage shares are over 70 percent. This implies that double- or triple-cropping was widespread in rural China in the 1930s, and the diffusion of minor commercial crops was still limited at that time.

From the examination of cropping variations, we expect to find that the more land each farmer possessed, the more diverse the cropping patterns. We estimate the coefficients of correlations between the number of crop items and the total cultivated lands by county. The distribution of the coefficient of correlations are depicted in Fig. 9.3. The coefficients are significant at the 1 percent level and positive in all counties except for Yuyao, where the coefficient is not significant even at the 10 percent level. Although the numerical values of the coefficients are differ among counties, they are uniformly dispersed from 0.3 to 0.7, and the level of those is relatively high in most counties. Thus, it appears that diversifications of cropping patterns and size of landholdings were positively correlated.

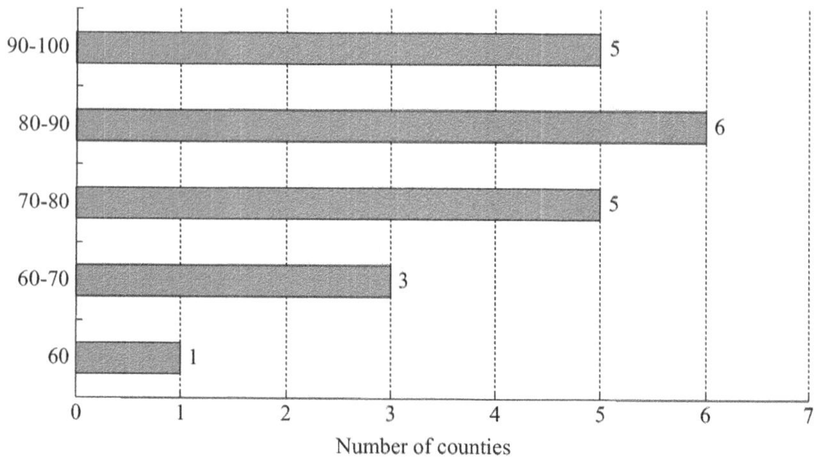

Fig. 9.2 Distribution of percentage share of cultivated areas of top three crops to total cultivated area

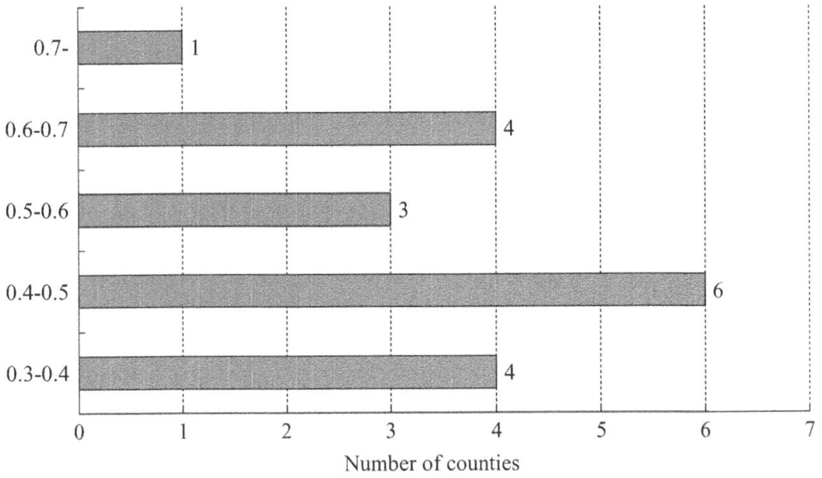

Fig. 9.3 Distribution of correlation coefficient between cultivated land area and number of crops cultivated

9.4 Controversy Around the Relationship between Farm Size and Crop Yield

In studies of agricultural economics in developing countries, it has been widely argued that the land productivity of agricultural products varies among landholding classes (Deolalikar 1981; Feder 1985). This relationship directly relates to the redistribution of land and the efficiency of tenant farming. Berry and Cline (1979) suggest that empirical studies on returns to scale of agriculture in developing countries are approximately constant. On the other hand, most existing studies identify an inverse relationship between farm size and yield per cropped areas. This relationship is due to the presence of a dual labor market where small farms face cheaper imputed labor cost, and seems to induce a higher labor/land ratio on small farms. Thus, small farmers achieve higher yield than large farmers (Feder 1985, pp. 297–298).[2]

[2] Feder (1985) also theorized that the inverse relationship in terms of the failure (underdevelopment) of the labor market was due to the supervision of labor and the failure of the financial market.

It should be noted that failures in the labor market do not necessarily induce this inverse relationship. Other possibilities could relate to failures in the input markets or some combination of the two. For example, small farmers were not able to easily obtain loans from the formal financial sector in 1929–1933 China because the formalizations under reconstruction had not yet fully developed and the introduction of credit societies was in its infancy (see Chaps. 2 and 10; Fu and Turvey 2018). Even in places where there were some developments collateral was generally insufficient. Consequently, credit constraints were widespread. In comparison, larger farmers might have had an advantage in acquiring capital to invest in agriculture to compensate for labor deficiency, and may have achieved higher productivity than small farmers. If true, this would indicate a positive relationship between landholdings and land productivity.

Huang (1985) and Cao (1996) provide two of the main examinations of the agricultural relationships of China for this period. The results of their studies are conflicting, however. Huang shows that clear relationships between land productivity and farm size were not observed in northern China, but small farmers were less profitable than middle and large farmers. On the other hand, Cao suggested that positive relationships were found in southern Jiangsu province. The discrepancies between those studies were partially due to the differences of economic conditions in each studied region.

According to Cao (1996), industrialization and commercialization were relatively developed in southern Jiangsu areas, and opportunities for off-farm work were more abundant than in northern China. Cao suggests three reasons for this positive relationship: the small size of farms, the deficiency of fertilizer and agricultural labor inputs, and the wide dispersion of small farms. Thus, the agricultural production of small farmers was less intensive than that of large farmers, and they produced a lower yield, despite the fact that their total labor inputs exceeded that of large farmers. Still, both studies agree that the hypothesis of an inverse relationship between cropped area and yield is not proven.

In comparison, Buck (1937, pp. 278–280) pointed out that a positive relationship between farm size and yield was not observed from the aggregated survey data. In this estimation, Buck converted weights of all agricultural products into grain-equivalent and treated the conversion rates of all crops, with the exception of cotton, silk, and fuel, as the same. However, these estimation methods are somewhat arbitrary and could give rise to aggregation bias. Thus, we have conducted more rigorous statistical tests to examine the relationship between farm size and yield of each crop using the newly digitized micro datasets.

9.5 Empirical Tests for an Inverse Relationship

In order to determine whether an inverse productivity relationship was prevalent in the Republican era we adopt a simple method using Box–Cox transformations on an independent variable. We consider simple production functions for each crop defined as $y = \alpha + \beta g(x) + \varepsilon$, where y is the amount of product and $g(x)$ indicates the cultivated areas. The Box–Cox transformation is of a flexible form with

$$g(x) = \frac{x^{\lambda} - 1}{\lambda} \tag{9.1}$$

$$\frac{\partial g(x)}{\partial x} = x^{\lambda-1} \tag{9.2}$$

and

$$\frac{\partial^2 g(x)}{\partial x^2} = (\lambda - 1)x^{\lambda-2}. \tag{9.3}$$

If the marginal product of y decreases with cultivated area then $\lim_{\lambda \to 0} \frac{\partial g(x)}{\partial x} = \frac{1}{x}$ and the transform $g(x) = \mathrm{Log}(x)$ can be applied so the crop production function would take the semi-log-linear form $y = \alpha + \beta \mathrm{Log}(x) + \varepsilon$. Likewise, as $\lim_{\lambda \to 1} \frac{\partial g(x)}{\partial x} = 0$, $g(x) = x$, and the function takes a simple linear form as $y = \alpha + \beta x + \varepsilon$. However, when $\lambda \neq 0$ and $\lambda \neq 1$ then the direction and curvature of the response function is determined by $\frac{\partial g(x)}{\partial x}$ and $\frac{\partial^2 g(x)}{\partial x^2}$. For example, when $0 < \lambda < 1$ response function is concave it will increase at a decreasing rate but when $\lambda > 1$ it will increase at an increasing rate.

We test the two kinds of single hypothesis ($\lambda = 0$ and $\lambda = 1$) using a likelihood ratio (LR) test. In case the null hypothesis $\lambda = 0$ is rejected and $\lambda = 1$ is not rejected, we can conclude that the function of land productivity takes a linear form. On the other hand, if $\lambda = 1$ is rejected and $\lambda = 0$ is not rejected, it would take a semi-log-linear form. If both hypotheses are rejected, the functional form would be $g(x) = \frac{x^{\lambda} - 1}{\lambda}$. Thus, the Box–Cox transformation and null hypotheses of our tests are summarized as follows:

$$g(x) = \begin{cases} \frac{x^{\lambda} - 1}{\lambda}, & \text{if } \lambda \neq 0 \\ \log x, & \text{if } \lambda = 0 \\ x & \text{if } \lambda = 1 \end{cases} \tag{9.4}$$

We concentrate our analysis on five crops, including the top three grain crops from those summarized in Table 9.4, as well as rapeseed and seed cotton. Table 9.4 complements Table 9.3, and reports the proportions of cultivated area of grain crops to total cultivated areas of grain. We note that the percentage shares of wheat and rice are strikingly high. For example, rice accounts for more than 60 percent of grain production in 11 counties. The share of wheat occupies a relatively lower share than that of rice, though the number of counties where wheat was raised is slightly higher than that of rice. Since the percentage share of cultivated areas of barley and kaoliang are lower than those of wheat and rice, the former seem to be raised as second or third grain crops in surveyed areas. On the other hand, the production of millet is concentrated in Gansu province (Ninghua and Wuwei), where the crop appears to be a staple food.

Except for grains, as shown in Table 9.4 the percentage share of legume, oil crops, and cotton were considerably high. These crops were not only important sources of protein for farmers, but also were major commercial products. Therefore, we focus on these three main grain (wheat, rice, and barley) and two non-grain crops (rapeseed and cotton) to examine the relationship between cultivated area and yield.

The relations between cultivated land and output of each farmer by crop and county are depicted in Figs. 9.4, 9.5, 9.6, 9.7, and 9.8. The horizontal axis measures the cultivated area and the vertical axis measures the amount of output.[3] All values of crop production and cultivated area are

[3]The county codes in these figures include Fengyang (1001), Fuyang (1002), Ho (1003), Wuhu (1004), Tangki (1101), Yuyao (1102), Nanyung (1201), Sinyang (1202), Kishui (1301), Yincheng (1302), Yunmeng (1303), Ninghua (1401), Wuwei (1402), Nanchang (1501), Pengste (1502), Tuchang (1503), Kwanyun (1601), Kunshan (1602), Wusih (1603), Yencheng (1604), Chungshan (1701), Ningwu (1801), Neikiang (1901), Ta (1902).

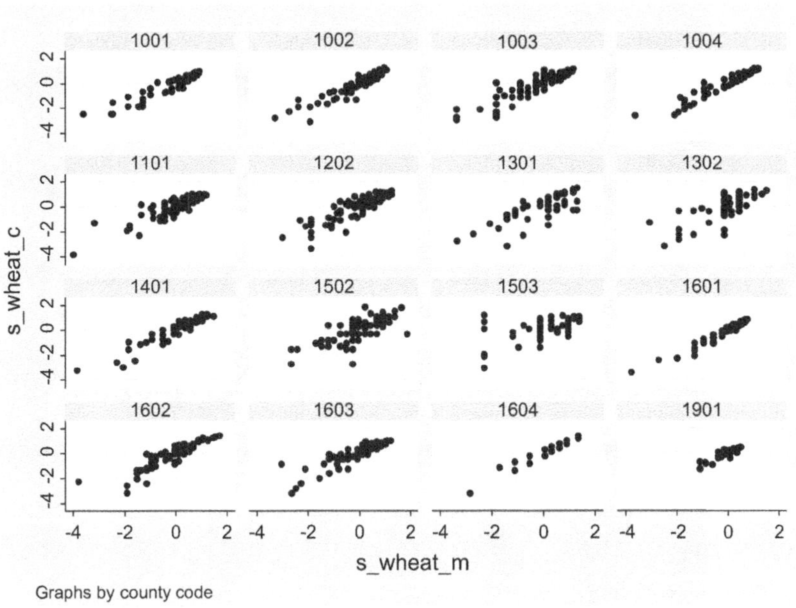

Fig. 9.4 Plot of wheat production and cultivated area (normalized)

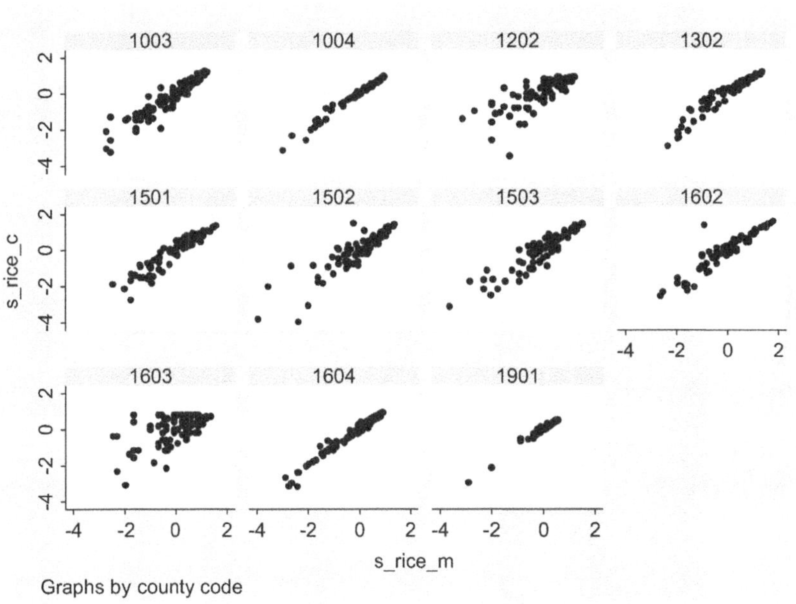

Fig. 9.5 Plot of rice production and cultivated area (normalized)

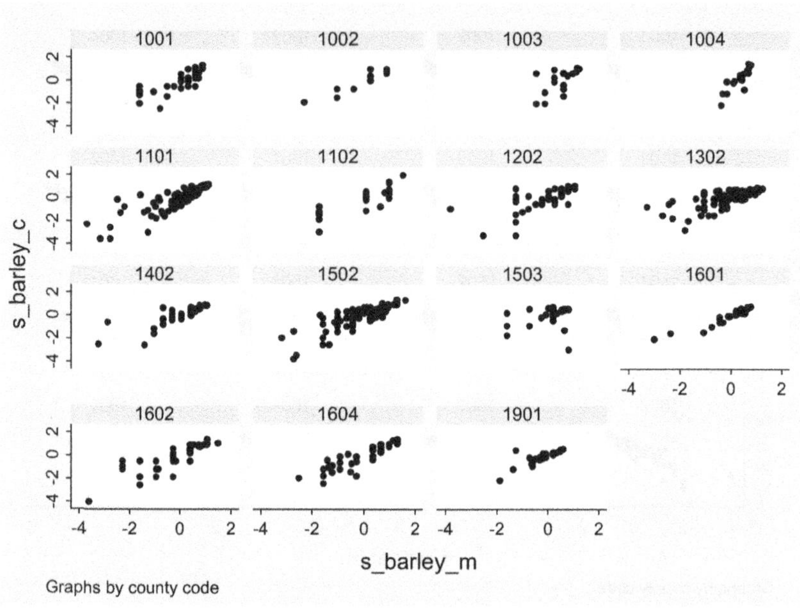

Fig. 9.6 Plot of barley production and cultivated area (normalized)

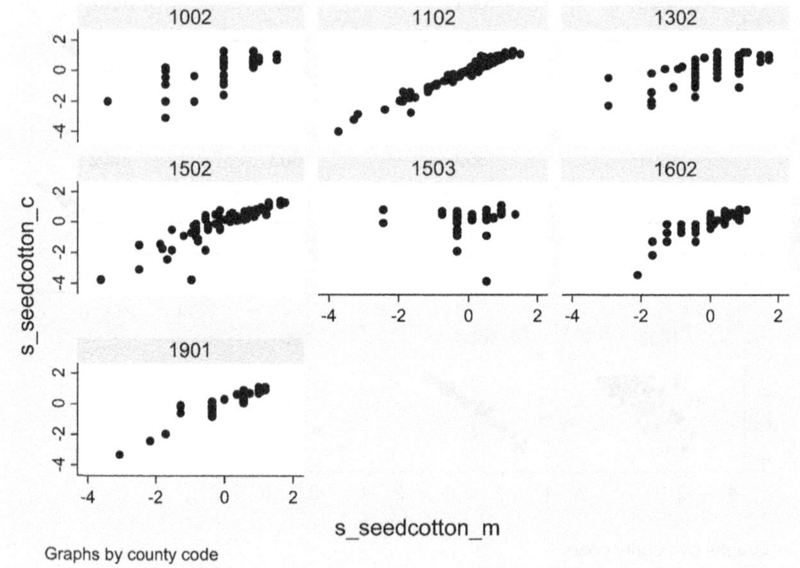

Fig. 9.7 Plot of seed cotton production and cultivated area (normalized)

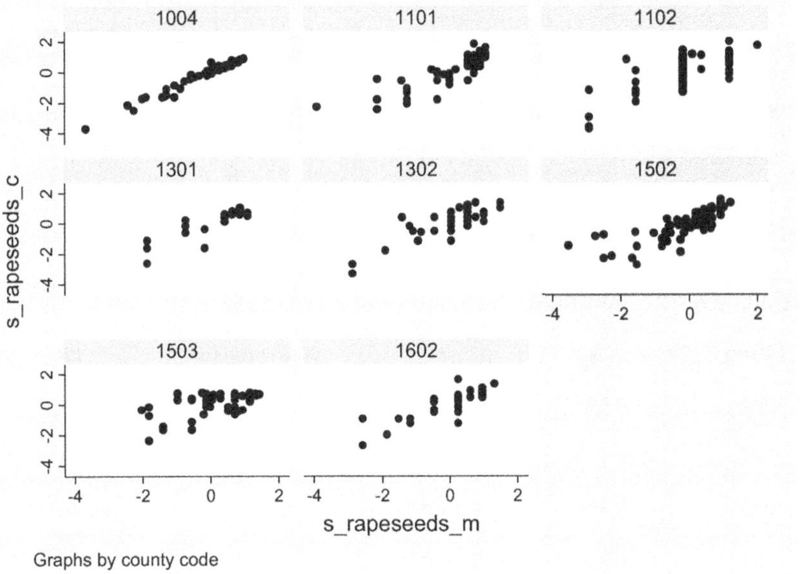

Fig. 9.8 Plot of rapeseed production and cultivated area (normalized)

standardized by county for comparison. From these figures, clear positive correlations between the amount of production and cultivated areas are observed in wheat and rice, while positive correlations are also found, but are not necessarily obvious, in other crops. In addition, the shapes of production function are relatively varied among crops and counties, especially the disparity among counties in non-staple crops such as rapeseed and seed cotton. This is partially due to small number of sample households who raised such crops.

In order to test the functional form of crop production, we apply a Box–Cox transformation by crop and county. The results of estimations are summarized in Table 9.5 and the details of the estimated results are also reported in Tables 9.6, 9.7, 9.8, 9.9, and 9.10.[4] Since the null hypothesis $\lambda = 0$ is rejected in almost all cases, we concentrate on the linearity test of $\lambda = 1$. Table 9.5 shows that the null hypothesis $\lambda = 1$ is mostly accepted with regard to wheat and barley; the case accounts for 12 out of 15 counties in

[4] Note: *** Significant at 1% level, ** significant at 5% level, and * significant at 10% level.

Table 9.5 Summary of Box–Cox transformation test

	Total	*λ = 1 rejected*	*λ = 1 not rejected*
Wheat	15	3	12
Rice	11	4	7
Barley	10	2	8
Rapeseed	7	3	4
Seed cotton	6	3	3

Note: Null hypothesis are tested at a 10% significant level

Table 9.6 Results of Box–Cox transformation test (wheat)

County	*λ*	z	*Null hypothesis*	*LR*
1001	0.843	10.090***	$\lambda = 0$	77.28***
			$\lambda = 1$	3.32*
1002	0.675	6.910***	$\lambda = 0$	38.55***
			$\lambda = 1$	10.09***
1003	0.897	9.130***	$\lambda = 0$	66.46***
			$\lambda = 1$	1.06
1004	0.840	8.150***	$\lambda = 0$	49.86***
			$\lambda = 1$	2.28
1101	0.979	7.400***	$\lambda = 0$	43.18***
			$\lambda = 1$	0.02
1202	1.008	6.050***	$\lambda = 0$	35.57***
			$\lambda = 1$	0.00
1301	1.409	4.180***	$\lambda = 0$	17.30***
			$\lambda = 1$	1.55
1302	0.841	3.700***	$\lambda = 0$	18.47***
			$\lambda = 1$	0.44
1401	0.978	10.420***	$\lambda = 0$	74.24***
			$\lambda = 1$	0.06
1502	1.180	4.470***	$\lambda = 0$	25.80***
			$\lambda = 1$	0.49
1503	0.964	1.420	$\lambda = 0$	2.07
			$\lambda = 1$	0.00
1601	0.887	19.510***	$\lambda = 0$	148.17***
			$\lambda = 1$	5.84**
1602	0.996	8.130***	$\lambda = 0$	54.19***
			$\lambda = 1$	0.00
1603	1.026	6.570***	$\lambda = 0$	37.86***
			$\lambda = 1$	0.03
1604	1.367	5.680***	$\lambda = 0$	21.27***
			$\lambda = 1$	2.30
1901	1.127	9.720***	$\lambda = 0$	71.27***
			$\lambda = 1$	1.24

Table 9.7 Results of Box–Cox transformation test (rice)

County	λ	z	*Null hypothesis*	*LR*
1003	0.876	10.900***	$\lambda = 0$	103.03***
			$\lambda = 1$	2.26
1004	1.111	29.740***	$\lambda = 0$	243.27***
			$\lambda = 1$	8.65**
1202	0.900	4.420***	$\lambda = 0$	20.54***
			$\lambda = 1$	0.23
1302	1.422	13.150***	$\lambda = 0$	142.51***
			$\lambda = 1$	16.97***
1501	1.122	8.910***	$\lambda = 0$	76.42***
			$\lambda = 1$	0.97
1502	0.877	6.560***	$\lambda = 0$	38.21***
			$\lambda = 1$	0.82
1503	0.827	7.960***	$\lambda = 0$	50.56***
			$\lambda = 1$	2.65
1602	1.187	8.700***	$\lambda = 0$	70.39***
			$\lambda = 1$	2.01
1603	1.130	2.890**	$\lambda = 0$	10.29***
			$\lambda = 1$	0.11
1604	1.093	26.040***	$\lambda = 0$	283.43***
			$\lambda = 1$	4.96**
1901	1.106	41.880***	$\lambda = 0$	273.61***
			$\lambda = 1$	15.33***

wheat and seven out of 11 counties in barley, while the hypothesis is rejected in about half of the counties concerning rapeseed and seed cotton.

We find that the value of significant λ tends to be less than 1 for wheat and barley in nine out of 15 counties and nine out of ten counties, respectively. This suggests that land productivity diminishes as the cultivated land area increases. By its flexible concave form in this range, the lower the value for λ, the more rapidly the farm-size output relationship diminishes. In comparison, λ for rapeseed and cotton seed tend to be greater than 1 for five out of seven counties for rapeseed and five out of six counties for seed cotton. These results suggest that land productivity for rapeseed and seed cotton increases with the cultivated area, and the rate of increase is greater as λ increases. Rice shows a mixed trend, with the hypothesis rejected in four of 11 counties, while the value of λ is greater than 1 for seven counties and less than 1 for four counties. Whether increasing size tends to diminish productivity or accelerate it depends on more localized conditions. Unfortunately, we cannot so easily determine what those conditions might be.

Table 9.8 Results of Box–Cox transformation test (barley)

County	λ	z	*Null hypothesis*	*LR*
1001	0.389	1.390	$\lambda = 0$	2.01
			$\lambda = 1$	3.88**
1002	0.899	2.540**	$\lambda = 0$	5.17**
			$\lambda = 1$	0.08
1003	0.726	0.910	$\lambda = 0$	0.99
			$\lambda = 1$	0.10
1004	1.024	1.430	$\lambda = 0$	3.08**
			$\lambda = 1$	0.00
1101	0.688	5.100***	$\lambda = 0$	25.79***
			$\lambda = 1$	4.60**
1102	0.796	2.300**	$\lambda = 0$	6.64*
			$\lambda = 1$	0.32
1202	0.452	1.400	$\lambda = 0$	1.90
			$\lambda = 1$	2.65
1302	0.782	3.780***	$\lambda = 0$	15.25***
			$\lambda = 1$	1.03
1402	0.749	4.390***	$\lambda = 0$	17.79***
			$\lambda = 1$	1.98
1502	1.074	4.750***	$\lambda = 0$	25.49***
			$\lambda = 1$	0.11
1503	2.539	1.510	$\lambda = 0$	4.53**
			$\lambda = 1$	1.58
1601	0.715	9.650***	$\lambda = 0$	49.79***
			$\lambda = 1$	11.79***
1602	0.725	3.750***	$\lambda = 0$	14.70***
			$\lambda = 1$	1.84
1604	0.601	1.920*	$\lambda = 0$	3.68*
			$\lambda = 1$	1.55
1901	0.990	5.600***	$\lambda = 0$	26.98***
			$\lambda = 1$	0.00

9.6 Summary and Conclusions

From our estimations, we can come to some general conclusions about the relationship between land and productivity. If we loosely (and quite arbitrarily) assume that $0.85 \leq \lambda < 1.15$ capture approximately linear relationships we find approximately 44 percent of crop/region combinations are approximately linear, that for $\lambda < 0.85$ about 33 percent have a positive relationship, but productivity decreases with size, while

Table 9.9 Results of Box–Cox transformation test (rapeseed)

County	λ	z	*Null hypothesis*	*LR*
1004	1.085	23.200***	$\lambda = 0$	197.49***
			$\lambda = 1$	3.31*
1101	0.380	1.990**	$\lambda = 0$	3.69**
			$\lambda = 1$	9.55**
1102	1.670	5.040***	$\lambda = 0$	25.26***
			$\lambda = 1$	4.49**
1301	1.195	2.400**	$\lambda = 0$	7.34*
			$\lambda = 1$	0.16
1302	1.751	3.320***	$\lambda = 0$	15.99***
			$\lambda = 1$	2.64
1502	0.229	1.490	$\lambda = 0$	2.27
			$\lambda = 1$	18.77***
1503	2.154	2.100**	$\lambda = 0$	6.41**
			$\lambda = 1$	1.68
1602	0.806	2.850**	$\lambda = 0$	8.80**
			$\lambda = 1$	0.44

Table 9.10 Results of Box–Cox transformation test (seed cotton)

County	λ	z	*Null hypothesis*	*LR*
1002	0.889	2.790**	$\lambda = 0$	10.69***
			$\lambda = 1$	0.12
1102	1.106	14.090***	$\lambda = 0$	119.30***
			$\lambda = 1$	1.82
1302	1.121	3.190***	$\lambda = 0$	10.74***
			$\lambda = 1$	0.12
1502	1.316	7.400***	$\lambda = 0$	54.16***
			$\lambda = 1$	3.31*
1503	−0.763	−0.900	$\lambda = 0$	0.77
			$\lambda = 1$	2.99*
1602	2.096	6.650***	$\lambda = 0$	54.89***
			$\lambda = 1$	15.50***
1901	1.333	7.650***	$\lambda = 0$	46.89***
			$\lambda = 1$	3.95**

the remaining 23 percent, with $\lambda \geq 1.15$, have a relationship that tends to increase as the farm size increases. In terms of specific crops, taking the simple averages of λ, we can conclude that the relationship between land productivity and cultivated areas of wheat (0.9483) and barley

(0.8766) are near linear and slightly diminishing. Rice (1.059) and seed cotton (1.013) are also linear with a very slight upward bend, while rapeseed (1.159) reveals a steeper increase. If the negative value of λ for seed cotton in Tuchang is excluded, however, the average λ for seed cotton increases to 1.309, which indicates efficiency increases with farm size.

Differences across crops (and regions) are likely caused by variations in labor intensity and the level of commercialization of each crop. The amount of labor necessary for the production of rapeseed and cotton seed per farm was more than that for wheat and barley, while that of rice was more than twice that for wheat and barley.[5] In addition, the levels of commercialization were differed considerably between grain crops and others. The percentage shares of commercialized products of rapeseed and seed cotton were 61 percent and 37 percent, while those of wheat and barley were 29 percent and 18 percent. That of rice was only 15 percent, although the percentage of non-commercialized rice used as payments for land tenure was 22 percent, the highest of all crops (Buck 1937, pp. 233–239). Accounting for this share, the level of commercialization of rice was actually over 30 percent.

While we note differences across crops and regions, we find on average that wheat and barley, while showing positive productivity gains with farm size, see diminishing productivity as farm size increases. Rice and seed cotton are also very close to linear but have a slight advantage with larger farm size. Rapeseed and seed cotton (when one negative value is removed) show a greater degree of specialization, with productivity generally increasing as farm size increases. The different results appear to relate to the labor intensity and level of commercialization of each crop.

From these results, we can tentatively conclude that the relationship between cultivated area and land productivity differed among crops, and that it is necessary to take into consideration the characteristics of each crop and the diversity of production patterns among landholding strata for evaluating this relationship.

[5]Buck (1937, pp. 301–303) showed that the average numbers of labor days required for rapeseed and cotton production were 48 and 53 days, while that of wheat, barley, and millet were 26, 40, and 40, respectively. The number of labor days required for non-grain crops was higher. However, the average number of days for rice production was 82 days, much more than for commercial crops.

References

Berry, R.A., and W.R. Cline. 1979. *Agrarian Structure and Productivity in Developing Countries: A Study Prepared for the International Labour Office Within the Framework of the World Employment Programme*. Baltimore: Johns Hopkins University Press.

Buck, John Lossing. 1937. *Land Utilization in China: A Study of 16786 Farmers in 168 Localities, and 38256 Farm Families in Twenty Two Provinces in China, 1929–1933*. Vol. 1. Shanghai: University of Nanjing.

Cao, Xingsui. 1996. *Jiu Zhongguo Sunan Nongjia Jingji Yanjiu [Traditional Chinese Farmer Economy in Southern Jiangsu]*. Beijing: Central Translation Publisher. (in Chinese).

Deolalikar, Anil. 1981. The Inverse Relationship Between Productivity and Farm Size: A Test Using Regional Data from India. *American Journal of Agricultural Economics* 63 (2): 275–279.

Feder, Gershon. 1985. The Relation Between Firm Size and Farm Productivity: The Role of Family Labor, Supervision and Credit Constraint. *Journal of Development Economics* 18 (2–3): 297–313.

Fu, H., and C.G. Turvey. 2018. *The Evolution of Agricultural Credit During China's Republican Era, 1912–1949*. Basingstoke: Palgrave Macmillan.

Huang, Philip. 1985. *The Farmer Economy and Social Change in North China*. Stanford, CA: Stanford University Press.

Kuribayashi, Sumio, ed. 2007. *Restoration of Farm Survey of Rural China in 1930s and Comparison with the Present Sampling Survey of Chinese Farms*. Kawagoe: Tokyo International University.

Myers, Ramon. 1976. Socioeconomic Change in Villages of Manchuria during the Ch'ing and Republican Periods: Some Preliminary Findings. *Modern Asian Studies* 10 (4): 591–620.

Rawski, Thomas. 1989. *Economic Growth in Prewar China*. Berkley: University of California Press.

Yanagisawa, Kazuya. 2000. *Kindai Chugoku niokeru Nouka Keiei to Tochi Shoyuu* [*Agricultural Household Management and Land Ownership in Modern China*]. Ochanomizushobo (In Japanese).

CHAPTER 10

The Relationship Between Farm Size and Land Productivity in Early Twentieth-Century China

Hao Hu and Minjie Yu

10.1 Introduction

The relationship between farm size and land productivity is one of the more controversial topics in the field of agricultural development. During the transition from agricultural to modern societies, the policy implication of "advantage of small farms" or "scale economy" has become one of the theoretical foundations of agricultural development, land distribution, and social justice. The study of this theory can be traced back to the early twentieth century with A. V. Chayanov's (1986, p. 32) assessment of Lenin's new economic policy, in which he pointed out that there was a negative relationship between farm area and efficiency in Russia. Studies on developing countries in Asia (Alvarez and Arias 2004), Europe (Barrett et al. 2010),

H. Hu (✉)
College of Economics and Management, Nanjing Agricultural University, Nanjing, China
e-mail: huhao@njau.edu.cn

M. Yu
Business School, Huzhou University, Huzhou, China

H. Hu et al. (eds.), *Chinese Agriculture in the 1930s*,
https://doi.org/10.1007/978-3-030-12688-9_10

and Africa (Sen 1966; Feder 1985) have also provided empirical support for a negative relationship between farm size and land productivity, and this has become a typical descriptor for productivity in traditional agriculture. As for the causes of this phenomenon, two schools of thought have arisen. The first is based on imperfect factor markets, and the other is statistical bias. The former, represented by Sen (1966), Feder (1985), and others, argues that imperfect labor markets are the ultimate cause of the negative relationship. While the abundance of household or familial labor may have been sufficient to meet the demands of cultivation initially, as farm sizes increased, the family labor-to-land area diminished. At some point family labor reached its limit, and labor productivity diminished as farm size increased. The shadow price—or wage rate—for labor would naturally increase, but the critical periods of labor demand would occur exactly when smaller farms would be least able to supply labor. Limitations in communication, transportation, and infrastructure would impede labor mobility from non-farm regions, and even without these issues the shadow price of labor was likely to fall short of the industrial wage rate. The latter argument, represented by Bhalla and Roy (1988), and Benjamin (1995), among others, suggest that the omission of land quality variables and statistical errors related to land contributed to this negative relationship.

In this chapter we examine the land–productivity relationship using Buck's data. From the modernist view of development economics Buck's data permits a glimpse into the productivity of an almost purely peasant economy, at a time when China's industrial base was far below that of Western countries. Rural industry was scant, and the country's infrastructure placed pressing limits on labor mobility. These two characteristics suggest that labor supply in rural areas was high, and that imperfections in the mobility of labor and capital suppressed whatever wage advantage off-farm or migrant labor might provide. Thus, the macro characteristics of 1929–1933 China provide a rare opportunity to examine the labor market imperfection hypothesis, while the scope of data available to us, including regional productivity differences to capture heterogeneity in land use, form, and quality, can take into consideration the statistical bias argument. Thus, traditional agriculture is an ideal object for research. More specifically, agriculture in early twentieth-century China did not experience excessive administrative interference, the agricultural economy was relatively free, and the free market was initially formed and developed appropriately (Benjamin and Brandt 1997). However, due to data availability, there are few empirical studies on Chinese agriculture of that time, and research was confined to small regions. For example, based on South Manchuria Railway Bureau survey data, Benjamin and Brandt (1995, 1997) looked at women's contribution to agriculture, and the relationship between land distribution and income distribution in early twentieth-century Manchuria. Based on first-hand investigations by some senior scholars, Hou Jianxin (2001) estimated the agricultural labor productivity in the central part of Hebei Province. To date, there have been few studies on the relationship between farm size and land productivity in early twentieth-century China as a whole. With Buck's micro-level household data now available to us, we can explore the issue of an inverse productivity–land relationship and the role that labor plays in this phenomenon in far more detail than previously.

10.2 Farm Size and Land Productivity in China

Does the traditional Chinese agriculture show a negative relationship between farm size and land productivity? If so, what are the reasons for this negative relationship? Are labor market imperfections involved? To answer these questions we use Buck's data to examine land productivity for farms of different scales in different areas, analyze the related reasons, and discuss the experience from the point of view of traditional agriculture. Exploring land productivity differences and related reasons in early twentieth-century China is not only useful for objectively understanding Buck's survey, but helpful for fully comprehending traditional agricultural productivity, which is instructive for those low-income countries that are still at the traditional agricultural stage. China is a large country, with varying climate, soil, and other natural conditions in different areas, leading to various planting structures, cultivation practices, production technologies, and population density. Consequently, agricultural production in different regions is largely heterogeneous, while being generally homogenous within each region. As Buck described it at the time "*the whole country was divided into north and south from the west to the east of Huai River, ..., which were wheat region and rice region respectively.*"[1] The wheat region included the spring wheat area, winter wheat–millet area, and winter wheat–kaoliang area, while the rice region included the Yangtze rice–wheat area, rice–tea area, Sichuan rice area, double-cropping rice area, and southwestern rice area.[2] Accordingly, we apply Buck's method of regional division as a reference, and divide China into eight agricultural areas.

[1] Buck, *Chinese Farm Economy* (the Chinese Version), p. 8.

[2] Buck, *Land Utilization in China* (the Chinese Version), Nanjing, 1937, pp. 8.

Table 10.1 Research sample

Areas	*No. of counties*	*No. of households*	*Sample counties (no. of households)*
Wheat region			
Spring wheat area	4	408	Kaolan (1) (100), Kaolan (2) (100), Ningsia (100), Sining (108)
Winter wheat–millet area	14	1400	Loyang (99), Lingpao (101), Fowping (101), Pingliang (100), Tienshui (100), Anyi (100), Lin (100), Pingting (102), Tsincheng (100), Chenan (98), Chowchih (100), Shang (99), Sunyi (100), Weinan (100)
Winter wheat–kaoliang area	11	1087	Fowyang (97), Linchang (100), Nanyang (100), Changli (2) (102), Chengting (100), Kwanyun (99), En (2) (100), Fushan (92), Hweimin (101), Laiyang (98), Tsimo (98)
Rice region			
Yangtze rice–wheat area	9	929	Fengyang (100), Hofei (100), Liuan (101), Wuhu (100), Tsaoyang (100), Yingcheng (100), Pengtsch (101), Kunshan (83), Yencheng (4) (144)
Rice–tea area	18	1798	Siuning (100), Linhai (100), Lishui (100), Tangki (100), Tunglu (1) (105), Tunglu (2) (101), Tungyang (82), Yungkia(100), Changtch (101), Chen (2) (100), Sinhwa (100), Wukang (104), Yiyang (101), Fowliang (103), Kaoan (100), Nanchang (100), Teian (100), Tuchang (101)
Sichuan rice area	5	458	Mien (100), Mienyang (88), Neikiang (100), Suining (70), Ta (100)
Double-cropping rice area	6	605	Yungning (98), Jung (106), Chungshan (100), Koyiu (100), Kukong (101), Mowming (100)
Southwestern rice area	4	412	Pinchwan (102), Tsuyung (106), Yuankiang (100), Yuki (104)

The research sample comprised several counties in each agricultural area, each of which consisted of about 100 households (Table 10.1). After excluding missing individual data, the total represents 7097 households in 71 counties.

In traditional agriculture, the family population is positively related to the farm area,[3] so farm size per capita is the most effective indicator of households' land capacity. We took farm size per capita as the standard measure, and divided the farms into small, small and median, median, median and

[3] Buck, *Land Utilization in China* (the Chinese Version), Nanjing, 1937, pp. 516–517.

large, and large farms. This is similar to Buck's approach in *Land Utilization in China*.[4] As shown in Table 10.2, the farm size per capita obviously differed, but the distribution has some common features, the most obvious of which was that small and medium-sized farms accounted for about half.

10.3 Relationship between Farm Size and Land Productivity

Productivity is generally applied to land productivity and measured by the output per unit area. According to Buck's rural survey, the majority of farms, regardless of size, had a diversified planting structure. They produced various crops, including grain crops such as wheat, sorghum, barley, millet, and economic crops such as fruits, vegetables, cotton, and herbs. The problem with agricultural diversity is that tabulating a metric for "productivity" per farm or land area is burdened by inexact equivalence. Buck got around this problem by developing a crop index per unit area based on the total yield of all crops on a given cultivated land area expressed as "equivalent amount of grain."[5] We retain this measure in our assessment.

10.4 Productivity of Different-Sized Farms

Based on Buck's household micro data, we calculated the output per unit area of each household in each county as a criterion to measure land productivity. We then obtained the average output per unit area of farms of different sizes, such as small farms, small and medium-sized farms, medium-sized farms, large and medium-sized farms, and large farms. As shown in Table 10.3, because of the limitations of soil, climate, and other natural conditions, and the diversified planting structure, productivity varied markedly from area to area. For example, the average output per unit area was quite low in the spring wheat area where the natural conditions were quite poor and crop varieties were single; while the average output per unit area were relatively high in double-cropping rice area and southwestern rice area where the multiple cropping indices were higher and the proportion of cash crops were higher.

[4] Taking the arithmetic average of farm area per capita in each area as a based, the 25% standard deviation as the group distance, the group with the median was the median farm. Meanwhile, there were small & medium-sized farmers and small farmers downward, and median & large and large farms upward.

[5] Buck, *Land Utilization in China: Statistics* (the Chinese Version), Nanjing, 1937, pp. 211.

Table 10.2 Households' farm size per capita and its distribution unit: Mou, %

Areas	*Average farm size per capita*	*Minimum farm size per capita*	*Maximum farm size per capita*	*The proportion of farms of different sizes*				
				Small	*Small and median*	*Median*	*Median and large*	*Large*
Wheat region								
Spring wheat area	2.72	0.17	27.83	10.05	39.22	23.28	14.95	12.5
Winter wheat–millet area	3.51	0.11	19.82	35.64	12.14	22.36	9	20.86
Winter wheat–kaoliang area	4.03	0.2	34.85	21.44	29.25	21.53	9.66	18.12
Rice region								
Yangtze rice–wheat area	2.85	0.14	19.5	21.21	27.88	18.73	13.35	18.83
Rice–tea area	2.69	0.06	21.57	19.91	29.31	21.86	12.12	16.8
Sichuan rice area	2.29	0.18	22.21	16.16	32.75	23.8	12.23	15.06
Double-cropping rice area	3.11	0.53	18.46	19.17	29.75	22.81	13.06	15.21
Southwestern rice area	1.65	0.12	5.22	23.06	28.4	17.48	11.89	19.17

Source: Buck's household micro data, organized and calculated by the authors

Table 10.3 Average output per unit area unit: Equivalent amount of grain (kg)/mou

Areas	*Average output per unit area of farms of different sizes*				
	Small	*Small and median*	*Median*	*Median and large*	*Large*
Wheat region					
Spring wheat area	73.4	89.55	83.52	67.57	56.27
Winter wheat–millet area	93.96	85.47	82.94	80.35	74.6
Winter wheat–kaoliang area	207.36	128.79	128.71	128.53	112.62
Rice region					
Yangtze rice–wheat area	193.84	169.56	162.17	165.27	163.68
Rice–tea area	355.93	216.89	203.48	185.3	154.26
Sichuan rice area	181.54	179.88	166.28	157.56	167.03
Double-cropping rice area	291.28	222.5	206.39	197.47	192.62
Southwestern rice area	595.42	508.88	395.77	355.51	249.81

Source: Buck's household micro data, organized and calculated by the authors

We then compared the differences of productivity within the agricultural areas so that such differences do not affect the results. We next analyzed the effect of cropping structures on productivity differences. From this, we observed that there was indeed a negative relationship between cultivated land per capita and land productivity in all eight agricultural areas. In other words, we can visually observe the "advantage of small farms," meaning that output per unit area of small farms was the highest, then small and median farms, and finally large farms.

10.5 Scale Elasticity of Agricultural Production in Different Areas

Scale elasticity of production is the most important economic indicator of the relationship between scale and productivity: the elasticity of scale of agricultural production reflects the relationship between cultivated land area and productivity. We examine productivity empirically using a Cobb–Douglas production function, $y_i = AL_i^{\alpha}\ell_i^{\beta}$ in log form, and calculate the scale elasticity of agricultural production in different areas. The general regression equation is

$$\log(y_i) = A + \alpha\log(L_i) + \beta\log(\ell_i) + e_i \tag{10.1}$$

Since "*traditional agriculture is a resource based agriculture, whose main input is land and labor*" (Wang 1998), here y_i is the total output (equivalent amount of grain, KG) of farm i, L_i is the cultivated land area (mou), and ℓ_i is the labor input, expressed by the total working hours (month) of the whole adult labor force, including family labor and hired labor. Our interest in scale refers to the relationship between land and labor. An economically convenient approach to measuring relative scale is by identifying the properties of the labor–land isoquant, that is, the two-dimensional curve that represents the combinations of land and labor for a given level of output. For our Cobb–Douglas representation, the labor-dependent form of the isoquant is given by $\ell = \left(\frac{\bar{y}}{A}\right)^{\frac{1}{\beta}} L^{-\frac{\alpha}{\beta}}$, with marginal rate of substitution $\frac{d\ell}{dL} = -\frac{\alpha}{\beta}\frac{\ell}{L}$ and elasticity of substitution, $\varepsilon = -\frac{\alpha}{\beta}$. It is more convenient to represent labor–land substitution in its elastic form because it is a relative measure that can be compared across size classes. Another measure of importance is production elasticity, which we define here as $\varepsilon_P = \alpha + \beta$. Because the Cobb–Douglas is a flexible homogenous function, when $\alpha + \beta = 1$ the firm, or regional, size-dependent agricultural economy in our context, exhibits constant returns to scale. In this state a doubling of all inputs (labor and land) will double output. Likewise, for $\alpha + \beta > 1$ the economy exhibits increasing returns to scale (output will more than double if inputs are doubled), and for $\alpha + \beta < 1$ it exhibits decreasing returns to scale. The marginal elasticities for land (α) and labor (β) can also be described using similar terminology. For example, the resulting coefficient α is the elasticity of scale, with larger values capturing the percentage change in output given a percentage change in land. If $\alpha > 1$ land productivity is said to be increasing in scale, when $\alpha = 1$ it is constant in scale, and when $\alpha < 1$ it is diminishing in scale. If the inverse productivity–land hypothesis is correct then α should decrease as farm size increases.

On the basis of controlling counties, we estimated the land elasticity of agricultural production in each area, and found that the scale elasticity is less than 1. As shown in Table 10.4, the scale elasticity of the eight areas was less than 1, and the figure was far from 1 except in the Sichuan rice area, which meant there was an obviously strong negative relationship between farm size and land productivity. Table 10.5 presents the labor elasticities, which vary widely by farm size and agricultural region. Table 10.6 provides the combined production elasticities.

Table 10.4 Scale elasticity of agricultural land

Areas	*Small*	*Small and median*	*Median*	*Median and large*	*Large*
Wheat region					
Spring wheat area	0.65	0.71	0.25	0.69	0.22
	(0.0028)	(0.0000)	(0.1732)	(0.0373)	(0.0783)
Winter wheat–millet area	0.76	1.03	0.97	1.05	0.82
	(0.0000)	(0.0000)	(0.0000)	(0.0000)	(0.0000)
Winter wheat–kaoliang area	0.75	0.80	1.22	1.08	0.95
	(0.0000)	(0.0000)	(0.0000)	(0.0000)	(0.0000)
Rice region					
Yangtze rice–wheat area	0.82	0.73	0.77	0.90	0.27
	(0.0000)	(0.0000)	(0.0001)	(0.0001)	(0.0834)
Rice–tea area	0.55	1.06	0.97	0.94	0.85
	(0.0000)	(0.0000)	(0.0000)	(0.0000)	(0.0000)
Sichuan rice area	1.05	1.02	0.61	0.85	0.62
	(0.0000)	(0.0000)	(0.0009)	(0.0041)	(0.0001)
Double-cropping rice area	0.80	0.92	1.13	0.89	0.91
	(0.0000)	(0.0000)	(0.0000)	(0.0000)	(0.0000)
Southwestern rice area	0.67	0.60	0.46	0.58	0.88
	(0.0000)	(0.0000)	(0.0034)	(0.0044)	(0.0000)

Note: The *P* values are in parentheses

The labor elasticities are, for the most part, negative. The median and large farms grouping in the spring wheat area is an exception. Here, adding more labor would tend to increase output. This may suggest that for whatever reason the labor supply in this particular market is finite, so the shadow price of labor is high. But the spring wheat region in general tends to have higher labor productivity than the other regions. Indeed, for most regions and farm sizes the labor productivity is quite low and even negative in some cases, although those areas exhibiting such elasticities are not statistically different from zero. Even so, this suggests that labor is marginally unproductive. For median size farms in the winter wheat–kaoliang area, it is statistically different from zero. In other words, labor is so unproductive that adding additional labor leads, statistically, to a decline in agricultural output. From these observations it appears that the productivity size argument is explained in part by not only imperfections in labor supply but also generally ineffective or unproductive labor. Comparing Tables 10.4 and 10.5, it seems that productivity is determined not so much by labor, but by the land itself. In other words, on a relative scale the marginal value product of land exceeded the marginal value product of labor.

Table 10.5 Scale elasticity of agricultural labor

Areas	*Small*	*Small and median*	*Median*	*Median and large*	*Large*
Wheat region					
Spring wheat area	0.50 (0.0001)	0.78 (0.0000)	0.87 (0.0000)	1.0342 (0.0004)	0.83 (0.0000)
Winter wheat–millet area	0.31 (0.0000)	0.26 (0.0000)	0.28 (0.0000)	0.28 (0.0000)	0.33 (0.0017)
Winter wheat–kaoliang area	0.38 (0.0000)	0.12 (0.1359)	−0.22 (0.0635)	−0.15 (0.3738)	−0.15 (0.2388)
Rice region					
Yangtze rice–wheat area	0.27 (0.0084)	0.40 (0.0003)	0.60 (0.0001)	0.32 (0.1203)	0.71 (0.0008)
Rice–tea area	0.57 (0.0000)	0.03 (0.5963)	0.18 (0.0306)	0.22 (0.0246)	0.29 (0.0002)
Sichuan rice area	−0.08 (0.4983)	0.05 (0.6589)	0.45 (0.0028)	−0.04 (0.8647)	0.80 (0.0000)
Double-cropping rice area	0.24 (0.0038)	0.14 (0.0330)	0.06 (0.4070)	0.12 (0.1853)	0.04 (0.5637)
Southwestern rice area	0.14 (0.1562)	0.24 (0.0411)	0.15 (0.3617)	0.05 (0.8051)	−0.07 (0.5594)

Table 10.6 Production elasticities

Areas	*Small*	*Small and median*	*Median*	*Median and large*	*Large*
Wheat region					
Spring wheat area	1.15	1.49	1.12	1.72	1.05
Winter wheat–millet area	1.07	1.29	1.25	1.33	1.15
Winter wheat–kaoliang area	1.13	1.02	1	0.93	0.8
Rice region					
Yangtze rice–wheat area	1.09	1.13	1.37	1.22	0.98
Rice–tea area	1.07	1.09	1.15	1.16	1.14
Sichuan rice area	0.97	1.07	1.06	0.81	1.42
Double-cropping rice area	1.04	1.06	1.19	1.01	0.95
Southwestern rice area	0.81	0.84	0.61	0.63	0.81

Perhaps more striking with regard to regional and crop differences are the measures of scale economies obtained by adding α and β. These are provided in Table 10.6. If $\alpha + \beta = 1$, the agricultural economy is in a constant return to scale regime, suggesting that doubling all inputs (land and labor) will double output. When $\alpha + \beta > 1$, the economy is exhibiting increasing returns to scale, so that in doubling inputs output will more than double. Likewise, for $\alpha + \beta < 1$ the economy is exhibiting decreasing returns to scale so that a doubling of inputs will increase productive output by less than 100%. The production elasticities as provided in Table 10.6 reveal that only one region exhibits constant returns to scale and that is the winter wheat–kaoliang area, and only for median size operations. The majority of region–size combinations, 29, exhibit increasing returns to scale. Of this group, the median and large farm grouping in the spring wheat area is the largest, with a production elasticity of 1.72. The spring wheat, winter wheat–millet and rice–tea areas show increasing returns of varying scales for all farm sizes. Of the 11 region–size combinations exhibiting decreasing returns to scale, the most significant is the southwestern rice area where all farm sizes exhibit decreasing returns, with the medium size group being the lowest, with a production elasticity of 0.61.

Perhaps most remarkable about the production elasticities is that with the exception of southwestern rice area, the pattern across farm sizes appears to take on an inverse-U shape. In other words, the most scale-efficient farms are not the smallest or largest farms but one of the medium size farm groups. This inverse U does not hold exactly for the Sichuan rice area, which is more serpent-like, with high returns to scale for the largest farms.

The labor–land substitution elasticities, measured by computing $\frac{\alpha}{\beta}$, are provided in Table 10.7. Again, these are relative rather than absolute measures and are interpreted as the percentage change in labor required for a 1% increase in land, holding output constant. The substitution elasticities vary widely across regions and farm type. Generally speaking, high elasticities are driven by low or decreasing returns to labor (i.e. low beta). For example, small farms in the spring wheat area would have to decrease labor by only 1.297% for an increase in land size of 1%, but large farms in the double-cropping rice area would have to decrease labor by 20.67% for every 1% increase in land in order to keep output constant along the isoquant. Some care must be taken in interpreting these higher values because in many instances the denominators, being very close to zero, were not statistically different from zero. What we cannot say with too much certainty is

Table 10.7 Marginal rates of substitution

Areas	*Small*	*Small and median*	*Median*	*Median and large*	*Large*
Wheat region					
Spring wheat area	−1.297	−0.909	−0.291	−0.670	−0.271
Winter wheat–millet area	−2.49	−3.994	−3.422	−3.80	−2.528
Winter wheat–kaoliang area	−2.015	−6.660	5.639	7.036	64.969
Rice region					
Yangtze rice–wheat area	−3.058	−1.287	−1.287	−2.793	−0.388
Rice–tea area	−0.967	−35.333	−5.379	−4.354	−2.944
Sichuan rice area	13.869	−20.658	−1.350	19.89	−0.781
Double-cropping rice area	−3.268	−6.370	−18.666	−7.455	−20.671
Southwestern rice area	−4.665	−2.460	−3.155	−10.730	11.607

whether large elasticities are reflecting a fairly ineffectual labor force, or an increase in the scale of land due to improved quality or technical efficiency.

Likewise, positive elasticities of substitution are most likely driven by negative scales in labor. In other words, if increasing labor in the general case reduced output, then to achieve the same level of output one would have to increase the amount of hours of ineffectual labor in order to keep yield constant. This is unlikely to be true in any case since increasing land and holding labor constant would dominate any strategy of hiring more ineffectual labor. With this, it is best to interpret the positive elasticities to be, at most, zero.

10.6 Elasticities on the Labor/Land Ratio

As a final metric we examine the labor-to-land ratio directly. The estimating equation takes the form $y_i = A\left(\frac{\ell_i}{L_i}\right)^{\gamma}$. The elasticity measure, γ, captures the percentage increase in output as the labor-to-land ratio increases (Table 10.8). This is a measure of proportions rather than absolutes. Again the elasticities show substantial strength in the spring wheat area, but do not exhibit the same strength in an inverse-U proposition as supported by the production elasticity measures. Measures that are negative or close to

Table 10.8 Labor-to-land elasticities

Areas	*Small*	*Small and median*	*Median*	*Median and large*	*Large*
Wheat region					
Spring wheat area	0.49(0.0001)	0.75(0.0000)	0.86(0.0000)	0.90(0.0037)	0.80(0.0000)
Winter wheat–millet area	0.27(0.0000)	0.20(0.0002)	0.23(0.0011)	0.10(0.3398)	0.23(0.0158)
Winter wheat–kaoliang area	0.34(0.0000)	0.17(0.0175)	−0.22(0.0191)	−0.09(0.5384)	0.06(0.5275)
Rice region					
Yangtze rice–wheat area	0.24(0.0134)	0.35(0.0009)	0.60(0.0002)	0.26(0.1968)	0.72(0.0000)
Rice–tea area	0.51(0.0000)	0.004(0.9461)	0.12(0.1191)	0.13(0.1387)	0.20(0.0033)
Sichuan rice area	−0.07(0.5109)	0.02(0.8194)	0.44(0.0029)	−0.01(0.9543)	0.55(0.0013)
Double-cropping rice area	0.23(0.0040)	0.12(0.0542)	0.03(0.6667)	0.12(0.1694)	0.07(0.2833)
Southwestern rice area	0.20(0.0371)	0.33(0.0013)	0.36(0.0141)	0.29(0.1330)	0.03(0.8189)

zero are generally not statistically different from zero, providing results similar to the previous discussion. The same exception on median size farms in the winter wheat–kaoliang area with $\gamma = -0.22$ holds with significance.

10.7 Representative Isoquants

Using the above information we depict in Fig. 10.1 four isoquants that illustrate in a general way the labor–land tradeoffs in Buck's China. The isoquants are derived using $\ell = \left(\frac{\overline{y}}{A}\right)^{\frac{1}{\beta}} L^{-\frac{\alpha}{\beta}}$. Isoquants are for small and large farms for spring wheat, Yangtze rice–wheat, rice–tea, and southwestern rice areas. Representative output is from Table 10.3. The farm size is based on Buck (1937, Table 9.2, Chap. 9, p. 269) after conversion to modern mou by a factor of six. Although the x-axis runs from 1–10, these are normalized for ease of presentation. The farm size pairings (small, large)

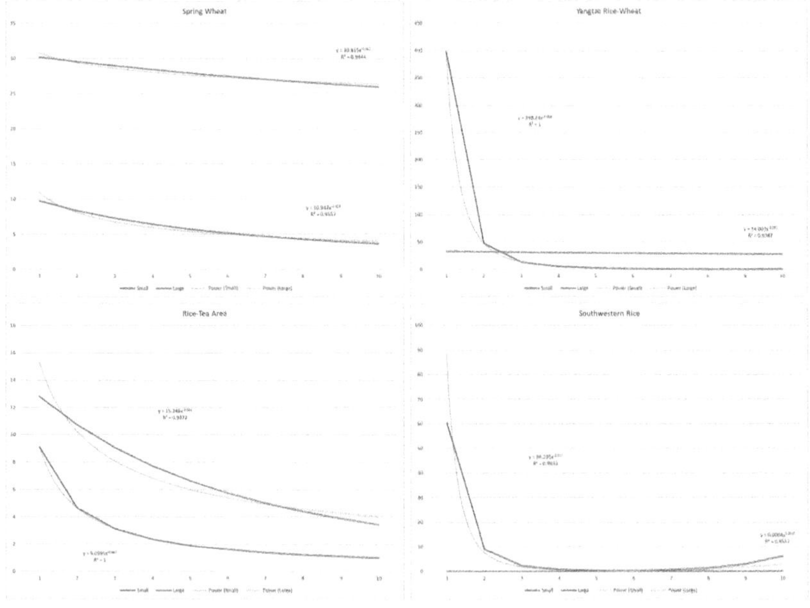

Fig. 10.1 Labor–land isoquants

are (13.8, 435), (4.8, 202.8), (3, 106.2), and (2.4, 61.2) for the spring wheat, Yangtze rice–wheat, rice–tea, and southwestern rice areas respectively.

Along with each isoquant we provide a power function estimate for which the exponent is an estimate of constant elasticity of substitution. These, of course, differ from those elasticities presented in Table 10.6, which implicitly assess the elasticities of substitution at the means of output and land. Nonetheless, the power estimates displayed in Fig. 10.1 provide an illustration of the relative differences in elasticity between the smallest and largest farms. The power elasticity pairs (small, large) are (−0.436, −0.067), (−3.058, −0.081), (−0.967, −0.582), and (−3.517, 3.8915). For the spring wheat, Yangtze rice–wheat, and rice–tea areas the elasticities of substitution of labor to land are higher for small farms than large ones. This is consistent with the results in, and discussion centered on, Table 10.7. We include the isoquants for the southwestern rice area to illustrate the degenerate case for large farms, with an (economically implausible) upward-sloping isoquant. This is likely due to errant data of some sort or another, but as can be seen in Table 10.6 the production elasticities for southwestern rice are substantially lower than those of farms in other regions, and all exhibit decreasing returns to scale. If taken literally the degeneracy suggests that when expanding land, the quality of the new land is so poor that substantially more labor is required to maintain production at the same level as the last mou cultivated. This interpretation should not be overlooked. In time, population growth leads to the subdivision of existing land amongst family members, and also to the cultivation of new land with increasingly diminished productivity. To counter this loss in natural productivity either more labor is required or some gains in human capital and the deployment of new technologies would be necessary. If China was in a high-level equilibrium trap, as suggested by Elvin (1973), the abundance of labor stymied innovations (Fu and Turvey 2018, Chapter 2). This could result, at least in the short run, in certain forms of economic degeneracy, as observed in these data. So on one scale it is entirely possible to observe what Huang defines as "involution" and a diminishing marginal productivity of labor, while on another scale observing gains in aggregate output to keep up with population increases, even if just at the level of subsistence. If the diminishing marginal product of labor is not offset by some technical innovation then an almost inevitable consequence is a below-equilibrium poverty trap (Nurkse 1952; Elvin 1973; Huang 1985).

Based on the above elasticities it is evident that labor productivity was, on average, quite low. In the spring wheat area for example, a 1% increase in labor increased productivity by only 0.63%. The Sichuan rice area had the highest scale elasticity of 0.92, implying that in that area a 1% increase in labor would increase output by 0.92%. These results show that in terms of labor productivity, after adjusting for land area, there was heterogeneity across the various regions in traditional agriculture in early twentieth-century China. The next task is to determine whether or not there were measureable differences in productivity that might reveal an "advantage of small farms."

10.8 Possible Explanations for Differences in Production Efficiency

In the previous sections we evaluated scale economies in agricultural production based on land and labor using a classical Cobb–Douglas production function. However, there may be many other factors that differentiated scale economies during this Republican era. In the following subsections we use Buck's data to explore a number of interrelationships between agricultural inputs and the five classes of farm size. These include cropping patterns and farm diversification, fertilizer input, planting structure, animal labor input, and labor. In doing so we recognize that labor productivity is highly endogenous; two laborers of equal skill and strength might have very different productivity outcomes if one had access to animal labor and fertilizer, while the other did not. Nonetheless, the availability of Buck's micro data provides some useful insights into production relationships that give a greater understanding of production efficiency in the 1930s.

10.8.1 The "Advantage of Small Farms" and Different Household Behavior

Compared with modern agriculture, "*habits or customs were the main approaches to allocate resources in traditional agriculture*" (Li Chenggui 1997). So, did farms of different sizes have the same habits or customs? In other words, did differences in land productivity result from variations in household production behavior?

Land, which is the most basic element in agricultural production, along with technical conditions, planting methods, and so on, determines the ideal scale of production to maintain a livelihood or create profits. Some scholars tried to estimate the minimum land requirement per capita in a certain period or region, including Northern China in the early twentieth century. For example, Chen Hansheng (1985, p. 150) estimated that a family of five needed 20–30 mou; Taylor, an American scholar, considered that a family of five in North China needed 25 mou; while Chen Zhongmin (cf. Jianxin Hou 2001) estimated they needed at least 20 mou to maintain a minimum living standard. These estimates clearly vary to some extent, and only relate to households in North China, but they basically reflect land requirements for survival. Combining these figures, together with the multiple cropping index (MCI), we still had reason to believe that small farms could face pressure to survive because of limited land resources.

10.8.2 Different MCI

Did small farms adjust their production behavior in order to cope with the pressures of survival? If yes, improving planting density was a good solution to the issue of limited land area. At that time, multiple cropping was quite popular in China. On average, multiple cropping was carried out in half of the surveyed area. Almost all crops in the second season were planted after winter crops had been harvested[6]: "Nationwide, the cropping index of smaller farms were slightly higher than the larger ones. The percentages of multiple cropping area for small farms and large farms were 53% and 43%, respectively. Compared to wheat region, this relationship was even more remarkable in rice region, where the growth season was longer."[7]

The multiple cropping index (MCI), derived by Buck measured the number of crops grown per household. Diversification might have been a risk strategy, but for smaller farms that consumed a larger portion of grown crops, the land was more likely put to economic crops, whereas larger farms required less land for in-farm consumption and might grow a greater amount of cash crops. Table 10.9 summarizes the MCI across farms of different sizes. What is observed is basically a negative relationship between the cultivated land area and the MCI. An exception is the double-cropping rice area. Therefore, a possible explanation for the "advantages of small farms"

[6] Buck, *Chinese Farm Economy* (the Chinese Version), p. 222.
[7] Buck, *Land Utilization in China* (the Chinese Version), p. 361.

Table 10.9 MCI

Areas	*Average MCI of farms of different sizes*				
	Small	*Small and median*	*Median*	*Median and large*	*Large*
Wheat region					
Spring wheat area	1.48	1.34	1.13	0.93	0.8
Winter wheat–millet area	1.28	1.26	1.22	1.18	1.15
Winter wheat–kaoliang area	1.49	1.4	1.36	1.34	1.37
Rice region					
Yangtze rice–wheat area	1.79	1.64	1.59	1.57	1.49
Rice–tea area	2.12	1.79	1.63	1.47	1.33
Sichuan rice area	1.77	1.7	1.6	1.45	1.49
Double-cropping rice area	1.8	1.8	1.75	1.69	1.83
Southwestern rice area	1.79	1.72	1.63	1.61	1.42

Source: Buck's household micro data, organized and calculated by the authors

was that smaller farms that were more vulnerable to food security threats alleviated survival pressure by multiple cropping. The MCI difference between small farms and large farms was one of the factors affecting their respective land productivity. For example, in the spring wheat area smaller farms were 48% more diversified than large farms, whose diversity was only 80% of a typical farm in the region.

10.8.3 Differences in Fertilizer Input

In addition to increasing the planting density, increasing fertilizer input was also an effective way to increase output to cope with the pressure of survival. In the early twentieth century, "a major obstacle to solving the problem of chemical fertilizer was traffic inconvenience," therefore, "the farm fertilizer had two sources, one was manure, and the other was night soil, which were important in maintaining soil fertility." "The amount of fertilizer used was too small at that time, and more fertilizer should be used to produce high output."[8] According to the farms' fertilizer inputs recorded in Buck's rural survey (Table 10.10), the amount of fertilizer input per unit area decreased with the increase of cultivated land area,

[8] Buck, *Chinese Farm Economy* (the Chinese Version), pp. 317–319.

Table 10.10 Amount of fertilizer input per unit area unit: kg/mou

Areas	*Amount of fertilizer input per unit area of farms of different sizes*				
	Small	*Small and median*	*Median*	*Median and large*	*Large*
Wheat region					
Spring wheat area	1847.56	2210.86	1797.17	1597.92	1155.93
Winter wheat–millet area	614.93	797.75	695.48	582.57	512.2
Winter wheat–kaoliang area	1188.32	1210.64	1160.39	1525.7	1139.09
Rice region					
Yangtze rice–wheat area	1653.69	1658.61	1779.99	1822.93	2065.96
Rice–tea area	1495.08	893.91	808.68	748.24	349.35
Sichuan rice area	3609.2	1929.88	1401.4	1289.5	1258.66
Double-cropping rice area[a]	1453.1	1334.58	992.3	887.74	498.23
Southwestern rice area	1539.77	1269.67	1086.68	1145.97	845.31

Source: Buck's household micro data, organized and calculated by the authors

Note: [a]Due to the loss of the original data of fertilizer input in the double-cropping rice area, here the number for farmyard manure has been replaced. Because the "use of artificial fertilizers was still only at the beginning period, farmyard manure, such as manure, night soil, ashes, bean cakes, were mainly used" (see Buck, ***Land Utilization in China*** (the Chinese version), pp. 232). This substitution is not completely accurate, but it can at least reflect the difference in the amount of fertilizer input

except for the Yangtze rice and wheat region. Therefore, another possible explanation for the "advantages of small farms" was that small farms which were seeking to increase output increased fertilizer input to maintain soil fertility, effectively promoting land productivity. This could simply mean that there were sufficient animal and human sources of fertilizer for small farmlands, and not enough for large lands. Chemical fertilizers, at that time, were rare, and even if available capital constraints, including access to credit, and transportation would have been limiting factors.

10.8.4 Different Planting Structures

If the planting structure affected land productivity across counties, might it affect productivity among farms within the county? In other words, if small farms chose to cultivate crops with a higher output per unit area, then the different planting structure would induce a "negative relationship." Farms were mainly confronted with the choice between grain crops and cash crops. Traditional agriculture was mainly self-sufficient, and small farms in particular needed to survive. Therefore, if market conditions for grain crops were poor, small farms preferred coarse grains whose yield were higher and more stable; between grain crops and cash crops, they preferred grain crops (Ahmed 1981). Most cash crops, which tended to command better prices, could be used to buy more grain crops, but because of the many uncertainties concerning market transactions and relative prices, small farms preferred grain crops. If market conditions were good, the result could be the exact opposite.

It should be noted that, as Huang and other scholars have said, the planting structure was mainly affected by climatic conditions and market environment, but were the influences of these constraints on small and large farms consistent? At present, we are not able to give clear data results. But some of the phenomena found in the data-organizing process may partially confirm the above inference; therefore, we provide some examples in the hope that scholars can discuss this in the future. In some counties in the spring wheat and winter wheat–kaoliang areas, whose market conditions were better, small farms preferred wheat and rice, which were more expensive and required more labor input. They could also barter wheat and rice, whose prices were higher, for coarse grains whose calories were higher, to improve their subsistence level. In the Sichuan rice area where the tax pressure was greater, farmers generally grew tobacco and opium, but small farmers also preferred to cultivate edible crops, such as vegetables and carrots, thereby increasing land productivity. Therefore, different cropping structures may induce a negative relationship between cultivated land area and land productivity to some extent.

10.8.5 Different Animal Labor Input

Animals, mainly engaged in pulling, grinding, carrying water, and other heavy work, played an important role in Chinese traditional agriculture. The larger the farm size, the more economical it was to use animal labor, so compared to small farms, the large farms obviously had more labor animals. However, there was a strong sense of community within counties at that time, which meant that large farms often lent such animals to small farms.[9] So, from the point of view of animal labor per unit area, small

[9] Buck, *Land Utilization in China* (the Chinese Version), p. 235.

Table 10.11 The amount of animal labor input per unit area unit: Equivalent animal unit/mou

Areas	*Amount of animal labor input per unit area of different-sized farms*				
	Small	*Small and median*	*Median*	*Median and large*	*Large*
Wheat region					
Spring wheat area	0.0445	0.0671	0.0794	0.071	0.0465
Winter wheat–millet area	0.0264	0.0336	0.0391	0.0341	0.0306
Winter wheat–kaoliang area	0.0489	0.0379	0.0343	0.0351	0.0313
Rice region					
Yangtze rice–wheat area	0.0544	0.0447	0.045	0.039	0.0351
Rice–tea area	0.0731	0.0694	0.0695	0.0575	0.0453
Sichuan rice area	0.0384	0.0547	0.0479	0.0491	0.0358
Double-cropping rice area	0.116	0.135	0.1013	0.0962	0.0537
Southwestern rice area	0.6435	0.562	0.3257	0.3336	0.1927

Source: Buck's household micro data, organized and calculated by the authors

farms may not have been at a disadvantage, and may even have been dominant. Buck translated cattle, buffalo, mule, horse, and other different types of animal labor into an equivalent animal labor unit. According to Table 10.11, except for the spring wheat area and winter wheat–millet area, small farms had more animal labor units per unit area. It is worth noting that animal labor did not completely substitute for human labor, because in traditional agriculture, animal labor mostly needed to be operated by humans. In that sense, animal labor was complementary to human labor to some degree. The more animal labor units per unit area, the more improved the land productivity; ultimately this led to the "advantage of small farms." Note, however, that this does not imply that small farms had more animals than large farms, but when measured against the size of land they had a greater proportion of such animals.

Also observed in Table 10.11 is an inverted-'U' relationship across farm sizes in the spring wheat, winter wheat–millet, Sichuan rice, and double-cropping rice areas. On balance, more intensive animal use was found for the small and median or median farm sizes, which may suggest that smaller farms faced more capital constraints or were more reliant on human labor. For the large farms, the results suggest that inasmuch as animal labor was fully utilized in other farm sizes, the excess slack in animal labor would somehow have to be filled with human labor in order to achieve the same level of output.

All the above is an analysis of the reasons why the "advantages of small farms" can be derived from different household production behavior. We believe that small farms faced greater pressure with regard to survival; therefore, in the process of agricultural production, small farms would increase planting density, use more fertilizer, adjust planting structures, and use more animal labor to ease this pressure. Although the four influencing factors discussed played different roles in different areas, the production behavior variations controlled by custom between small and large farms triggered a negative relationship between farm size and land productivity.

10.8.6 "Advantages of Small Farms" and Labor Markets

As for factor allocations in traditional Chinese agricultural, labor had a strong elasticity of supply, whereas land and capital had a certain pattern of rigidity. In this unbalanced supply structure, the labor factor with higher elasticity of supply relaxed the supply restrictions of other material factors to a certain extent. Thus, an increase in agricultural output was achieved by adding more labor input (Li Chenggui 1997), while other inputs, mostly related to labor, required little external inputs (Wang 1998). The difference in household production behavior described was essentially a difference in labor input. What was the reason for the difference in agricultural labor input between small and large farms? We have to shift the focus away from agricultural households to the external factor markets.

The traditional agricultural factors market was mainly represented by the land and labor markets. If the rural factor markets developed well, when the household could not absorb all the family labor due to scarce land, it would offset land restrictions by selling labor or renting land, and vice versa. However, when the land leasing market and the labor market were not perfect, the households with land scarcity had to use more labor per unit area, characterized by intensive and meticulous farming such as higher MCI. There was a quite small elasticity of land supply in early twentieth-century China. An empirical study by Benjamin and Brandt in Manchuria confirmed that any change of land ownership or in land lease relationships was very slow, so the labor market had become the most important channel to balance the agricultural resource endowment

(Benjamin and Brandt 1997). For example, hired labor could make up for a shortage in the family labor force, while seasonal or long-term off-farm employment could reduce household labor surplus. From Buck's data, at that time, 15% of agricultural work was done by the hired labor;[10] for 13.3% of agricultural households at least one family member had a non-agricultural job, for 36.1% of agricultural households at least one family member held a part-time job.[11] It was reasonable to believe that China had already formed a labor market to a certain extent. If the labor market developed well, the surplus rural labor force would be effectively transferred; in comparison, the labor force remaining in rural areas had to continue to work in agricultural production and improve land productivity through various approaches.

In the late nineteenth century, when agricultural capitalism was in crisis in America and Europe, Chayanov demonstrated the necessity of small farms pushing out capitalist farms, based on the theory of "small farms' self-exploitation of labor" (Wen 2011). This theory, in fact, illustrated how the mechanism of labor market imperfections affected productivity, and Sen concluded it was a "cheap labor" hypothesis: In traditional agriculture, the difference of labor opportunity cost between small and large farms was the fundamental reason for using a different density of labor. Specifically, small farms were dominated by family laborers, whose rational choice was to increase labor input until the marginal output of labor was zero. That is to say, the opportunity cost of the labor force of small farms was very low, and may even have been zero, which was consistent with the basic fact that there were less non-agricultural employment opportunities in traditional agriculture. Meanwhile, we have shown that small farms faced much pressure for survival, so workers' leisure value was relatively low. In contrast, large farms, with more of the characteristics of a business organization, relied more on hired labor. Therefore, after the marginal output of labor equaled the level of wages, hired labor would no longer be used. In addition, the hiring process itself required searching, hiring, and supervision costs, so the large farms' opportunity costs of labor input were relatively high. The difference in opportunity costs resulting from the imperfect labor market led to small farms using more intensive labor input.

Addressing this theory, we compared hired labor (Table 10.12) and labor input (Table 10.13). From the data in Table 10.12, it can be seen

[10] Buck, *Land Utilization in China* (the Chinese Version), p. 305.
[11] Buck, *Land Utilization in China* (the Chinese Version), p. 303.

Table 10.12 Employment of hire labor unit: %

Areas	*Percentage of hired labor in total labor of farms of different sizes*				
	Small	*Small and median*	*Median*	*Median and large*	*Large*
Wheat region					
Spring wheat area	0.92	4.69	11.56	11.81	12.5
Winter wheat–millet area	3.06	3.54	7.39	8.48	15.59
Winter wheat–kaoliang area	4.39	4.49	9.91	17.76	35.98
Rice region					
Yangtze rice–wheat area	3.95	8.24	14.62	13.79	20.16
Rice–tea area	10.31	7.65	11.13	11.81	15.47
Sichuan rice area	5.51	11.52	13.77	17.69	29.64
Double-cropping rice area	8.21	9.94	9.08	10.59	13.3
Southwestern rice area	5.28	7.79	6.46	6.86	8.38

Source: Buck's household micro data, organized and calculated by the authors

Table 10.13 Labor input unit: Working hours of equi-adult labor/mou

Areas	*Labor input of farms of different sizes*				
	Small	*Small and median*	*Median*	*Median and large*	*Large*
Wheat region					
Spring wheat area	6.24	3.12	1.68	1.2	0.8
Winter wheat–millet area	2.58	1.4	1.06	0.87	0.65
Winter wheat–kaoliang area	2.91	1.51	1	0.8	0.6
Rice region					
Yangtze rice–wheat area	4.61	2.53	1.73	1.44	1.06
Rice–tea area	5.09	2.16	1.56	1.37	0.96
Sichuan rice area	5	2.8	2.06	1.62	1.17
Double-cropping rice area	3.91	2.32	1.83	1.55	1.03
Southwestern rice area	5.58	3.62	2.66	2.46	1.65

Source: Buck's household micro data, organized and calculated by the authors

that although farms in different areas relied on hired labor to varying degrees, on the whole, compared with small farms, large farms depended more on such labor. Table 10.13 showed the working months of adult (including family and hired) laborers in farms of different sizes. Obviously, with the increase of cultivated land area per capita, the labor input per unit area decreased. These two tables reflect the hypothesis of "labor self-exploitation" and "cheap labor." They show that the labor opportunity cost of small farms, who mainly depended on family labor due to the imperfect labor market, was smaller than that of large farms, or even zero; as a result, small farms input their large amount of surplus labor into limited land resources by maximizing their livelihood rather than pursuing the best allocation efficiency. Intensive labor input increased output per unit area, thereby increasing land productivity. On the contrary, the large farms, who needed more hired labor, could not achieve such an effective allocation of resources subject to the imperfect labor market.

Therefore, the heterogeneity of rural households' production behavior essentially relied on labor input differences due to variations in opportunity costs resulting from the imperfect labor market. In other words, the different labor input affected the production behavior and ultimately led to negative relationship between farm size and land productivity.

10.9 Conclusion

In summary, in early twentieth-century China, although there were differences in cultivated land area, planting structure, and so on, an overall negative relationship between farm size and land productivity was present. This concept of the "advantages of small farms" conformed to the idea of traditional agriculture. However, we cannot explain this negative relationship only in terms of differences in households' production behavior or features of labor market. It is possible for us to gain a more comprehensive understanding of Chinese traditional agriculture by considering both of these and clarifying their mechanisms. This chapter has argued that the imperfect labor market was the root cause of the "advantages of small farms," and that different households' production behavior was its external performance and directly led to a negative relationship.

Finally, we tried to compare the results of this study to research on farm size and productivity in China today (Chen et al. 2011; Benjamin and Brandt 2002). Some studies had shown that China's agriculture still had a very weak negative relationship between the farm area and productivity, but the influence of imperfect labor market was declining. The development of agriculture in China is very complicated, but the development of factor markets, including the labor market, is undoubtedly a necessary condition for changing the small-scale peasant economy into a modern agricultural economy.

Appendix: Detailed Production Coefficients

Table A1 Production coefficients, spring wheat area

	Coefficient	*SE*	t-*statistic*	P
Small				
C	4.4120	0.2082	21.1943	0.0000
B_1	0.6541	0.2046	3.1972	0.0028
B_2	0.5041	0.1171	4.3035	0.0001
Small and median				
C	4.3818	0.3052	14.3568	0.0000
B_1	0.7055	0.1621	4.3516	0.0000
B_2	0.7764	0.1129	6.8741	0.0000
Median				
C	5.6735	0.4592	12.3544	0.0000
B_1	0.2541	0.1851	1.3728	0.1732
B_2	0.8729	0.1465	5.9599	0.0000
Median and large				
C	4.0870	0.8594	4.7557	0.0000
B_1	0.6893	0.3234	2.1314	0.0373
B_2	1.0342	0.2769	3.7347	0.0004
Large				
C	6.0029	0.4093	14.6669	0.0000
B_1	0.2245	0.1248	1.7988	0.0783
B_2	0.8280	0.1264	6.5521	0.0000

Table A2 Production coefficients, winter wheat–millet area

	Coefficient	*SE*	t-*statistic*	P
Small				
C	4.7997	0.1101	43.5753	0.0000
B_1	0.7622	0.0580	13.1496	0.0000
B_2	0.3054	0.0547	5.5820	0.0000
Small and median				
C	4.1903	0.1626	25.7702	0.0000

(*continued*)

Table A2 (continued)

	Coefficient	*SE*	t-*statistic*	P
B_1	1.0285	0.0674	15.2493	0.0000
B_2	0.2575	0.0528	4.8803	0.0000
Median				
C	4.2839	0.2174	19.7055	0.0000
B_1	0.9649	0.0802	12.0302	0.0000
B_2	0.2820	0.0680	4.1478	0.0000
Median and large				
C	3.9626	0.2940	13.4796	0.0000
B_1	1.0519	0.1035	10.1609	0.0000
B_2	0.2768	0.1057	2.6188	0.0095
Large				
C	4.6415	0.3028	15.3301	0.0000
B_1	0.8216	0.0974	8.4393	0.0000
B_2	0.3250	0.1022	3.1798	0.0017

Table A3 Production coefficients, winter wheat–kaoliang area

	Coefficient	*SE*	t-*statistic*	P
Small				
C	5.4948	0.1473	37.2936	0.0000
B_1	0.7571	0.0916	8.2690	0.0000
B_2	0.3757	0.0761	4.9391	0.0000
Small and median				
C	5.1795	0.1707	30.3437	0.0000
B_1	0.7952	0.0754	10.5534	0.0000
B_2	0.1194	0.0798	1.4953	0.1359
Median				
C	4.1537	0.2474	16.7843	0.0000
B_1	1.2175	0.0929	13.1027	0.0000
B_2	−0.2159	0.1157	−1.8652	0.0635
Median and large				
C	4.5838	0.4204	10.9046	0.0000
B_1	1.0843	0.1487	7.2911	0.0000
B_2	−0.1541	0.1726	−0.8933	0.3738
Large				
C	4.9766	0.3124	15.9284	0.0000
B_1	0.9479	0.0999	9.4926	0.0000
B_2	−0.1459	0.1234	−1.1817	0.2388

Table A4 Production coefficients, Yangtze rice–wheat area

	Coefficient	*SE*	t-*statistic*	P
Small				
C	5.2395	0.1464	35.7986	0.0000
B_1	0.8152	0.1100	7.4088	0.0000
B_2	0.2666	0.1000	2.6648	0.0084
Small and median				
C	5.2956	0.2202	24.0533	0.0000
B_1	0.7341	0.1159	6.3332	0.0000
B_2	0.4020	0.1085	3.7054	0.0003
Median				
C	5.0212	0.4238	11.8487	0.0000
B_1	0.7677	0.1859	4.1300	0.0001
B_2	0.5965	0.1530	3.8986	0.0001
Median and large				
C	4.8852	0.5580	8.7555	0.0000
B_1	0.8967	0.2269	3.9520	0.0001
B_2	0.3211	0.2052	1.5646	0.1203
Large				
C	6.5262	0.4182	15.6039	0.0000
B_1	0.2765	0.1588	1.7413	0.0834
B_2	0.7135	0.2101	3.3968	0.0008

Table A5 Production coefficients, rice–tea area

	Coefficient	*SE*	t-*statistic*	P
Small				
C	6.0993	0.0752	81.1190	0.0000
B_1	0.5495	0.0557	9.8614	0.0000
B_2	0.5683	0.0548	10.3687	0.0000
Small and median				
C	5.1068	0.1343	38.0349	0.0000
B_1	1.0600	0.0665	15.9312	0.0000
B_2	0.0300	0.0566	0.5301	0.5963
Median				
C	5.1897	0.2041	25.4277	0.0000
B_1	0.9655	0.0886	10.8985	0.0000
B_2	0.1795	0.0827	2.1700	0.0306
Median and large				
C	5.1715	0.2220	23.2991	0.0000
B_1	0.9361	0.0910	10.2836	0.0000
B_2	0.2150	0.0950	2.2633	0.0246

(*continued*)

Table A5 (continued)

	Coefficient	*SE*	t-*statistic*	P
Large				
C	5.2394	0.1974	26.5358	0.0000
B_1	0.8506	0.0702	12.1245	0.0000
B_2	0.2889	0.0762	3.7910	0.0002

Table A6 Production coefficients, Sichuan rice area

	Coefficient	*SE*	t-*statistic*	P
Small				
C	4.9679	0.1671	29.7367	0.0000
B_1	1.0513	0.1390	7.5609	0.0000
B_2	−0.0768	0.1128	−0.6807	0.4983
Small and median				
C	4.9952	0.2132	23.4244	0.0000
B_1	1.0205	0.1210	8.4357	0.0000
B_2	0.0494	0.1117	0.4423	0.6589
Median				
C	5.6410	0.3690	15.2868	0.0000
B_1	0.6068	0.1772	3.4245	0.0009
B_2	0.4495	0.1466	3.0652	0.0028
Median and large				
C	5.4310	0.6529	8.3181	0.0000
B_1	0.8513	0.2837	3.0005	0.0041
B_2	−0.0428	0.2497	−0.1713	0.8647
Large				
C	5.4276	0.4018	13.5093	0.0000
B_1	0.6221	0.1498	4.1519	0.0001
B_2	0.7965	0.1556	5.1196	0.0000

Table A7 Production coefficients, double-cropping rice area

	Coefficient	*SE*	t-*statistic*	P
Small				
C	5.8088	0.1535	37.8322	0.0000
B_1	0.7965	0.0913	8.7233	0.0000
B_2	0.2437	0.0825	2.9541	0.0038

(*continued*)

Table A7 (continued)

	Coefficient	*SE*	t-*statistic*	P
Small and median				
C	5.4589	0.1584	34.4609	0.0000
B_1	0.9167	0.0758	12.0918	0.0000
B_2	0.1439	0.0670	2.1490	0.0330
Median				
C	4.8816	0.1947	25.0668	0.0000
B_1	1.1330	0.0864	13.1201	0.0000
B_2	0.0607	0.0730	0.8318	0.4070
Median and large				
C	5.4841	0.2569	21.3501	0.0000
B_1	0.8864	0.1016	8.7239	0.0000
B_2	0.1189	0.0890	1.3367	0.1853
Large				
C	5.5042	0.2043	26.9421	0.0000
B_1	0.9116	0.0704	12.9531	0.0000
B_2	0.0441	0.0761	0.5796	0.5637

Table A8 Production coefficients, southwestern rice area

	Coefficient	*SE*	t-*statistic*	P
Small				
C	6.6756	0.1391	47.9996	0.0000
B_1	0.6704	0.1158	5.7872	0.0000
B_2	0.1437	0.1005	1.4297	0.1562
Small and median				
C	6.7574	0.1798	37.5847	0.0000
B_1	0.5985	0.1103	5.4249	0.0000
B_2	0.2433	0.1178	2.0659	0.0411
Median				
C	6.9907	0.2959	23.6251	0.0000
B_1	0.4613	0.1519	3.0370	0.0034
B_2	0.1462	0.1592	0.9183	0.3617
Median and large				
C	6.7155	0.3885	17.2842	0.0000
B_1	0.5837	0.1951	2.9916	0.0044
B_2	0.0544	0.2163	0.2482	0.8051
Large				
C	5.8548	0.2993	19.5590	0.0000
B_1	0.8821	0.1269	6.9501	0.0000
B_2	−0.0760	0.1296	−0.5864	0.5594

References

Ahmed, I. 1981. Farm Size and Labor Use: Some Alternative Explanations. *Oxford Bulletin of Economics and Statistics* 43 (1): 73–88.

Alvarez, A., and C. Arias. 2004. Technical Efficiency and Farm Size: A Conditional Analysis. *Agricultural Economics* 30 (3): 241–250.

Barrett, C.B., M.F. Bellemare, and J.Y. Hou. 2010. Reconsidering Conventional Explanations of the Inverse Productivity–Size Relationship. *World Development* 38 (1): 88–97.

Benjamin, D. 1995. Can Unobserved land Quality Explain the Inverse Productivity Relationship? *Journal of Development Economics* 46 (1): 51–84.

Benjamin, D., and L. Brandt. 1995. Markets, Discrimination and the Economic Contribution of Women in China: Historical Evidence. *Economic Development and Cultural Change* 1995: 63–104.

———. 1997. Land, Factor Markets, and Inequality in Rural China: Historical Evidence. *Explorations in Economic History* 34 (2): 460–494.

———. 2002. Property Rights, Labor Markets, and Efficiency in a Transition Economy: the Case of Rural China. *Canadian Journal of Economics* 35 (4): 689–716.

Bhalla, S.S., and P. Roy. 1988. Mis-specification in Farm Productivity Analysis: The Role of Land Quality. *Oxford Economic Papers* 40: 55–73.

Chayanov, A.V. 1986. *The Theory of Peasant Economy*. Madison: University of Wisconsin Press.

Chen, Hansheng. 1985. *The Chinese Peasants*. Shanghai: Fudan Press.

Chen, Z., W.E. Huffman, and S. Rozelle. 2011. Inverse Relationship Between Productivity and Farm Size: the Case of China. *Contemporary Economic Policy* 29 (4): 580–592.

Elvin, M. 1973. *The Pattern of the Chinese Past: A Social and Economic Interpretation*. Palo Alto, CA: Stanford University Press.

Feder, G. 1985. The Relation Between Farm Size and Farm Productivity: the Role of Family Labor, Supervision and Credit Constraints. *Journal of Development Economics* 18 (2–3): 297–313.

Fu, H., and C.G. Turvey. 2018. *The Evolution of Agricultural Credit During China's Republican Era, 1912–1949*. Cham: Palgrave Macmillan.

Hou, Jianxin. 2001. Agricultural Labor Productivity in Central Part of Hebei Province in early 20th Century. *Chinese Agriculture History* 1: 57–67.

Huang, P. 1985. *The Peasant Economy and Social Change in North China*. Stanford, CA: Stanford University Press.

Li, Chenggui. 1997. Resource Allocation and Efficiency Level of Traditional Agriculture. *Chinese Agriculture History* 1: 74–83.

Nurkse, R. 1952. Some International Aspects of the Problem of Economic Development. *The American Economic Review* 42 (2): 571–583.

Sen, A.K. 1966. Peasants and Dualism with or Without Surplus Labor. *Journal of Political Economy* 74 (5): 425–450.

Wang, Siming. 1998. Conditions and Constraints: Resources, Technology, Institutions and Culture—An Analytical Framework for the Study of Agricultural Development. *Chinese Agriculture History* 1: 73–80.

Wen, Liming. 2011. The Failure of Agricultural Capitalism in the Western Countries in the Late Twentieth Century and Early Nineteenth Century. *Journal of Suzhou University of Science and Technology* 7: 59–65.

CHAPTER 11

Farm Credit Demand and Supply in 1930s China

Calum G. Turvey and Hong Fu

11.1 Introduction

In this chapter we put Buck's recovered data on farm credit to the econometric test.[1] The objective is to recover as best as possible the economic relationship between credit demand and interest rates, by specifying two endogenous equations for credit supply and credit demand. The interest rates and the quantities of credit observed represent an agreement between the borrower and lender and, as such, are simultaneously determined. We investigate non-productive or consumption loans and also production loans,

[1] In the modern era we have undertaken some efforts to understand credit demand and supply conditions. See, for example, Kumar et al. (2013), Turvey et al. (2012), Turvey and Kong (2010), Turvey et al. (2010), Verteramo-Chiu et al. (2014), and citations therein.

C. G. Turvey (✉)
Cornell University, Ithaca, NY, USA
e-mail: cgt6@cornell.edu

H. Fu
School of Insurance, Shandong University of Finance and Economics, Jinan, China

H. Hu et al. (eds.), *Chinese Agriculture in the 1930s*,
https://doi.org/10.1007/978-3-030-12688-9_11

while taking into consideration possible endogeneity between credit and productivity as well. We define a demand curve as quantity dependent with interest rates as independent variables, and the supply curve as interest-rate dependent, with loan quantity as an independent variable. In our three-stage least squares (3SLS) structure, loan amounts and interest rates are treated as instrumental variables when used as explanatory variables. We consider interest rates, agricultural productivity, and special farm expenditures such as weddings and funerals as drivers of demand, and loan amount, productivity, crop yield risk, and source of credit as drivers of supply. Productivity is linked to credit amount and interest rates as well as farm size, productive and market animals, hired labor, and so on as productivity factors.

Our findings are rather interesting. The first is we find no statistical difference between special expenditures on weddings, funerals, birthdays, birth of sons, and dowries between farmers who borrow and those that do not. Thus, against the conventional wisdom of the day we cannot attribute debt to special expenditures as a matter of course. Of course, it happens in some cases, and many farmers with such expenditures might have liked to obtain credit but could not, but we cannot make a broad sweeping statement to the effect that special expenditures were a strong source of credit.

A second interesting result was that farmers who did borrow had an almost perfectly inelastic demand for credit. The most reliable of our results suggest that this nearly perfectly inelastic demand revealed little or no sensitivity to the actual interest rate. Farmers were interest-rate takers. The lender on the other hand acted as a local monopolist of sorts, willfully increasing the interest rate as the amount of credit demanded increased. But we also find, across a number of different types of lenders, no significant difference in the interest rate charged. Interest rates were about 3% per month or 36% per year, and there was clearly heterogeneity in rates set by suppliers, but by and large no one group could be singled out as overtly exploitive. Either all lenders were usurers—or none were usurers—depending on whether one considers 3% or higher per month usury. Importantly, this finding applies to friends and relatives. In other words, there was no statistical difference in the rates on offer from a friend or relative, a business, or landlord or other type of lender.

11.2 Summary of Buck's Rediscovered Data Used in Demand Analysis

Table 11.1 summarizes the data available to the analysis, the number of observations recorded with valid data, and summary statistics across all observations. Every observation measures each household and the number of observations for each variable category such as interest rate ($N = 3044$)

Table 11.1 Descriptive statistics of Buck's data used in econometric credit demand analyses

Variable names	*Observation numbers* (N)	*Mean*	*Std. dev.*	*Min.*	*Max.*
Dummy loan (0/1) (Chinese currency)	3044	0.30	0.46	0	1.00
Total loan (Chinese currency)	3044	28.65	108.10	0	2000.00
Total interest rate (%)	3044	1.08	2.64	0	50.00
Production (average production/ year* *mou*)	2024	20.74	17.53	0	220.66
Weighted interest rate (%)	3044	1.05	2.53	0	50.00
Weighted interest rate for indebted farmers (%)	840	3.82	3.57	0.33	50.00
Average interest rate (%)	3044	0.54	1.32	0	25.00
Average interest rate for indebted farmers (%)	839	1.95	1.88	0.25	25.00
Consumptive loan (Chinese currency)	3044	21.54	90.63	0	2000.00
Consumptive interest rate (%)	3044	0.88	2.47	0	50.00
Productive loan (Chinese currency)	3044	7.11	49.34	0	1710.00
Productive interest rate (%)	3044	0.19	0.85	0	8.00
Farm area square	2125	4121.59	13,130.38	0	256,036.00
Farm square	2125	39.25	50.82	0	506.00
Labor animal	2127	1.60	2.21	0	24.00
Productive animal	2127	0.46	1.25	0	21.08
Relatives (0/1)	3044	0.05	0.21	0	1.00
Friends (0/1)	3044	0.04	0.20	0	1.00
Informal businesses (0/1)	3044	0.11	0.31	0	3.00
Formal/semi-formal businesses (0/1)	3044	0.01	0.10	0	1.00
All other borrowing sources (0/1)	3044	0.07	0.25	0	1.00
Wedding	4317	69.45	166.81	0	3500.00
Dowry	4319	19.84	76.09	0	1500.00
Birthday	4319	2.13	31.30	0	1500.00
Birth of son	4319	2.83	20.56	0	500.00
Funerals	4319	52.91	203.23	0	7000.00
Other special expenditures	4318	5.53	155.65	0	9800.00
Income from other sources (%)	1824	1.48	6.06	0	50.00
Yield risk (average production/ highest production)	2020	0.68	0.18	0	3.88
If hired labor (0/1)	1824	0.57	0.50	0	1.00
If subsidiary labor (0/1)	1824	0.62	0.49	0	1.00
Year 1929 first half	4421	0.2721104	0.4450968	0	1.00
Year 1929 second half	4421	0.317349	0.4654972	0	1.00
Year 1930 first half	4421	0.3168966	0.4653194	0	1.00
Year 1930 second half	4421	0.3155395	0.4647829	0	1.00

(*continued*)

Table 11.1 (continued)

Variable names	*Observation numbers (N)*	*Mean*	*Std. dev.*	*Min.*	*Max.*
Year 1931 first half	4421	0.2972178	0.4570849	0	1.00
Year 1931 second half	4421	0.2065144	0.4048497	0	1.00
Year 1932 first half	4421	0.2741461	0.4461334	0	1.00
Year 1932 second half	4421	0.2741461	0.4461334	0	1.00
Year 1933	4421	0.0226193	0.1487033	0	0.00
Hebei	4421	0.1164895	0.3208473	0	1.00
Shanxi	4421	0.1590138	0.3657303	0	1.00
Liaoning	4421	0.0239765	0.1529931	0	1.00
Jiangsu	4421	0.0622031	0.2415514	0	1.00
Zhejiang	4421	0.0454648	0.2083449	0	1.00
Anhui	4421	0.065596	0.2476026	0	1.00
Fujian	4421	0.0242027	0.1536953	0	1.00
Shandong	4421	0.1142276	0.3181234	0	1.00
Henan	4421	0.0228455	0.1494277	0	1.00
Guangdong	4421	0.0452386	0.2078506	0	1.00
Guangxi	4421	0.0461434	0.2098193	0	1.00
Sichuan	4421	0.0226193	0.1487033	0	1.00
Guizhou	4421	0.0689889	0.253464	0	1.00
Yunnan	4421	0.0454648	0.2083449	0	1.00
Shaanxi	4421	0.0226193	0.1487033	0	1.00
Gansu	4421	0.0452386	0.2078506	0	1.00
Qinghai	4421	0.0244289	0.1543939	0	1.00
Ningxia	4421	0.0226193	0.1487033	0	1.00
Suiyuan	4421	0.0226193	0.1487033	0	1.00

and farm area ($N = 2125$) differs due to the actual constraints of missing data. Thus, although we had 3044 households with valid data (including zero) on how much was borrowed, we had only 2125 observations with valid measures of land area. Including area in a regression immediately reduced the available sample to no more than 2125. Complicating matters even further, the actual number of observations available to run regressions is no greater than the numbers of valid and available data of all variables employed. The available data includes observations on credit and indebtedness, proportion of all farms and subsidiary works performed by family and hired labor by men, women, and children, amount and distribution of livestock, farm area devoted to different uses grouped by size of farm, number of mou of crop area devoted to variables crops, amount of fertilizer produced on the farm, yields per mou of all crops, savings, special expenditures, able-bodied men (over 15 and under 60 years of age), most frequent yield per mou of the byproduct of important crops, and amount and kinds of fertilizers applied per mou (Table 11.1).

11.3 Amount and Character of Farm Credit

As mentioned, our sample base is drawn from Buck's original sampling using the rediscovered household data (see Chaps. 2 and 4). Our data includes a total of 4421 individual household records. Of these we have valid data on 3044 households, with valid entries (non-empty) for credit use. Of these 3044 households a total of 902 or 29.6% of farm households reported debt of one kind or another (Table 11.2). Of these, 732 households (81.2% of borrowers and 24% of all 3044 households queried) borrowed for consumption purposes which includes special expenditures such as weddings, funerals and so on. Only 220 households, or 24.3% of all borrowers and 7.2% of sampled households, borrowed for production purposes. These suggest that only 50 households, or 5.54% of borrowers, actually borrowed for both production and consumption purposes. Buck reported that about 39% of households in China had some form of debt, with about 12% borrowing for productive purposes and 33% for unproductive purposes. So our sample is slightly biased with a smaller percentage of households borrowing for any cause; nevertheless, the proportions of our sample are reasonably consistent with Buck's assessment. We do not rule out differences in counting loans between our approach and Buck's as a source of discrepancy (percent of loans versus percent of households having loans).

On average, including zero amounts, the average farmer in our sample borrowed \$28.65, with a maximum of \$2000. For farms having credit the loan amounts averaged \$96.67, with those having consumption loans averaging \$89.55 and those with production loans \$98.39 (Table 11.2). In comparison, Buck found that indebted farmers owed about \$80 on average, with about \$19 borrowed for productive purposes and \$61 for unproductive or consumption purposes). Our average of \$96.67 exceeds Buck's of \$80 by \$16.67, which is a substantial margin. Again, the difference is due to calculation. Buck included zero values for production or

Table 11.2 Loan amount summary

Variable	*Obs.*	*Mean*	*Std. dev.*	*Min.*	*Max.*
Total loan	3044	28.64592	108.1006	0	2000
Dummy loan	3044	0.2963206	0.4567091	0	1
Total loan for indebted farmers	902	96.67205	181.3378	0.5	2000
Consumptive loan for indebted farmers	732	89.55321	167.5994	0.5	2000
Productive loan for indebted farmers	220	98.38745	157.4922	2	1710

consumption loans so long as the other had a non-zero value. The numbers we report exclude zero values which would be blended into Buck's averages. For example only 24.3% of borrowers borrowed for production. Multiplying this by $98.39 results in a loan amount of $23.91 which is close to Buck's $19. Likewise, with 81.2% of borrowers having average non-productive borrowings of $89.55 the weighted equivalent is $72.71, which is higher than the $61 reported in Buck. The not insignificant difference is likely due to our double counting of borrowers who had both production and consumption loans, and who account for about 50 farmers.

The data included interest rates for each loan made. In Table 11.3 (and all that follow) we report the weighted interest rate for indebted farmers (%), excluding records if this equals zero. We find that a simple average of (non-zero) interest rates charged on loans is 1.95% per month ($N = 839$). This treats each loan individually and independently of all others. But when the interest rates are weighted across all loans reported by farmers the weighted average interest rate is 3.82% ($N = 840$). The discrepancy can only be explained if the larger loans made to farmers come with significantly higher interest rates, a conclusion that becomes abundantly clear in our econometric analysis. The interest rates are quoted on a monthly basis so a 3.82% weighted rate implies a simple annual rate of 45.84%, which is in the range of interest rates reported elsewhere in the Chinese literature. Although there were some differences in interest rates charged, the differences in Buck's data do not appear to be great. The average rate of interest on productive loans was about 2.6% and higher for unproductive purposes. In some areas, such as the spring wheat area, there was a significant difference between the two, with production loans having a 2.9% rate (per month) and unproductive loans having a 3.7% rate. Again, differences in measurement explain why our rates, measured at the household level, differ from Buck's at the loan level. For example, suppose a farmer has a $10 loan at 2% and $90 loan at 4%. Buck's average would record an interest rate of 3%. In comparison, our weighted approach would be 2% * 0.10 + 4% * 0.90 or 3.8% as a weighted average for the farm.

Table 11.3 Interest rate summary

Variable	*Obs.*	*Mean*	*Std. dev.*	*Min.*	*Max.*
Weighted interest rate	3044	1.053162	2.534817	0	50
Weighted interest rate for indebted farmers	902	3.554131	3.577975	0	50
Consumptive interest rate for indebted farmers	732	3.676025	3.881945	0	50
Productive interest rate for indebted farmers	220	2.659318	1.833931	0	8

However, this comes with a wide range, from 0.33% to 50% per month. The upper range seems unreasonably high. On investigation we note that 50% interest rate was applied to a $5 loan in Wuwei, Gansu province. In fact, in Wuwei, there were 18 loans above 10% per month (11%–50%). While 10% per month is still substantially higher than average, there is enough documentary evidence elsewhere to suggest that this was not unusual at that time. However, this might be true of Gansu because it was very remote. Furthermore, Wuwei was surveyed from January to December 1932, which means that it is possible that interest rates were for loans made in 1931 while Gansu was still recovering from drought and famine (see Chap. 3). During this period, it is possible that some usurers took advantage of the farmers' plight and charged extraordinary rates. After considering the nature of data, we decided to keep what might appear to be outlier rates of interest. The high usury rates accounted for 18/680 of consumptive/non-productive loans, or 2.6% of the total. We also believe those high values in the consumption side of Buck's data in 1937–1937 would have been double-checked.

11.4 Sources of Farm Credit

In most cases, farmers obtained credit locally and personally. Buck queried farmers on where they sourced credit (Table 11.4) and found these included the same village (10%), relatives and friends, (39%), wealthy persons (6%), merchants (3%), farmers (5%), mortgagors (1%), shops(3%), in and/or near cities (5%), landlords (2%), neighbors (8%), adjacent villages (4%), others (7%), and unknown (7%). Unfortunately, Buck is quite unclear as to the nature of these loans. For example, assuming that most people in a village know each other, how are the 10% of borrowings in the same village different from friends or relatives or farmers; or how do wealthy farmers differ from mortgagors? In this case we know that in many cases the wealthy person would issue a mortgage in the form of a usufruct

Table 11.4 Rescaled borrowing sources of indebted farmers

Relatives	*17.85%*
Friends	14.29%
Informal businesses	39.29%
Formal/semi-formal businesses	3.6%
All other borrowing sources	25%

loan, but wealthy persons and landlords have also been identified as usurers. We also don't know the extent of *hui* in these villages or whether lending amongst friends and relatives or neighbors or others in the village is among contributing members of a rotating savings and credit organization (ROSCO). Since none are identified explicitly as credit cooperatives, banks, or native banks we don't know precisely what the nature of these loans, or lenders, are. Buck does admit that certain groups can be amalgamated and that is what we do. We define relatives, friends, informal businesses, formal/semi-formal businesses, and all other borrowing sources as distinct categories with the relative percentages in Table 11.4. Ultimately we use these as 0–1 binary variables in the supply regressions we run. The supply equation has the weighted interest rates as dependent variables, while including the credit-source variables as control variables will allow a determination of whether the supplier of credit differentially applies interest rates.

11.5 Uses of Credit and Special Expenditures

Our demand equation has as the dependent variable the amount of loan taken as a function of interest rates and the uses of the funds. As previously discussed, credit was used for production and non-productive purposes. Of the non-productive purposes we do not know how much was borrowed for food and other necessities of life, but Buck did collect data on special expenditures. The averages available in our sample ($N = 4319$) are provided in Table 11.5.

In our numbers we include all households even if they have zero special expenditures. Clearly, from an econometric point of view, where special expenditures are a (possible) driver of farm credit, even if a borrowing household had no special expenditures, the debt must then be assigned to productive purposes, which we observe, or non-special expenditures on

Table 11.5 Special expenditures of household sample

	N	*Average (including 0)*	*Standard deviation*	*Min.*	*Max.*
Wedding	4317	69.45	166.81	0	3500.00
Dowry	4319	19.84	76.09	0	1500.00
Birthday	4319	2.13	31.30	0	1500.00
Birth of son	4319	2.83	20.56	0	500.00
Funerals	4319	52.91	203.23	0	7000.00
Other special expenditures	4318	5.53	155.65	0	9800.00

food, shelter, clothing, health, or education, which we do not observe. This is important. Throughout the Republican era, nearly all of the literature discusses the sometimes ruinous impacts of borrowing for special expenditures on farm indebtedness and persistent poverty. Most important are the social effects of "saving face." This was discussed by Smith in *Chinese Characteristics* and evidence exists that it is a practice which persists to this day.

Buck also parses out special expenditures according to farm size and shows that there is an inverse relationship between some special expenditures, particularly weddings and funerals, and farm size. In other words, smaller and poorer farmers spend proportionately higher amounts on special expenditures most likely to improve social status. Ultimately we will show that this holds true, but in a discriminating way. With special expenditures, the social pressures are so critical and saving face so engrained in Chinese culture that, at least for weddings and funerals, the demand for cash is almost perfectly inelastic. This means that the lender can discriminate and charge higher interest rates for increased borrowing amounts—the more is borrowed, the higher the interest rate charged!

11.6 Livestock

We also include in our assessment a variable to capture whether or not the farm has livestock. Livestock can be classified as labor animals (e.g. donkeys) and productive animals. Labor animals (e.g. cattle, oxen) are raised for plowing, cultivation, and transportation, while productive animals are raised for household consumption (meat, eggs) or to be sold in the village market. Working livestock can also be an indicator of wealth, with larger farmers more than likely to require animal power, while smaller farmers use family or, if affordable, hired labor.

Table 11.6 summarizes the livestock holdings of our farmers. Only 2127 households had valid records for livestock, with the remainder having no entry. We believe that the missing observations were not collected, rather

Table 11.6 Livestock holdings of sample farms ($N = 2127$)

	N	*Average (including zero)*	*Standard deviation*	*Min.*	*Max.*
Labor animal	2127	1.60	2.21	0	24.00
Productive animal	2127	0.46	1.25	0	21.08

than being zero values, so treated these as such. However, we find on average that of these farmers the average number of labor animals they had was 1.60 per household, with only 0.46 productive animals per household. In most cases the production animals were small livestock, including poultry.

There appears to be a regional distribution of working animals dominated by water buffalo, oxen, donkey, mules, and horses for production, and hogs and sheep for food. Farms also kept chickens and other poultry, but in smaller numbers. Across China 25% of farm animals were productive animals. Buck found about 1.34 animal units per farm, which is similar to our sample of 1.60 animals per farm.

11.7 Size of Farm Business

We find that on average farmers in our sample had 39.25 mou of land (about 6.5 acres), with the largest being 506 mou. Only 2125 of farm households in our sample had valid entries for land. As a point of comparison, Buck reports China-wide farm size of about 3.8 acres, which is lower than our sample. Landholdings vary across China, with 5.1 to 7.3 acres in the wheat region areas and 2 to 3.5 acres in the rice-growing areas.

In our model we argue that credit demand, credit supply, and agricultural productivity are endogenous to each other. Households with greater productivity might have greater savings and less need for non-productive loans, or a larger demand for production loans. Larger farms with more land would have more collateral and therefore might be able to access credit at lower rates from creditors. In terms of simultaneous ordering it may also be the case that farms are larger because they could, for one reason or another, access credit. We also include in our regressions an additional term that squares the number of mou. This addition is included to gain some idea of the relationship between productivity and landholdings and economies of size. It is possible that given the limited resources available, farming larger tracts of land results in diminishing returns, with labor or productive animal units constraining the amount of land that could be efficiently cultivated.

11.8 Agricultural Productivity

Our measure of agricultural productivity is based upon the total weight of crop harvested on the farm. This measure was also used by Buck and we were constrained by our yield data that was presented in units of a jin, which is approximately 500 g or about 1 lb. The farms themselves were very diversified and many crops might have been grown by only one farmer in a county. The bulk of the weight harvested would come from the common crops grown by most farmers in a county.

Following Buck, we created a productivity measure that sums the total weight of crop harvested by each farm. This includes not only grains such as wheat, sorghum, rice but also vegetables and fruit. This presents two problems. The first, as mentioned, is the commingling of crops into a single measure. The second is the heterogeneity in crop type, not only between villages, but also provinces and agricultural regions. For the first, we recognize the intra-farm disparities that might result, but note also that in most cases the crops that dominated the county and region also dominated the productivity measure. On the second, we do include additional 1–0 binary variables to account for regional differences in dominant cropping patterns. From our data there were 2024 households with production data. All data were converted from a local weight measure to the common jin measure of 500 g and all land was converted from local mou to common mou. We find that on average farms produced 20.74 jin/mou with a standard deviation of 17.53 jin/mou and a maximum of 220.66 jin/mou.

11.9 Yield Risk

We have written at length in Chap. 3 on the many calamities faced by Chinese farmers, but how these calamities translate into crop yield losses at the farm level is not easily calculable. Buck, however, made an attempt at capturing yield risk by asking farmers about their highest yield in memory, which could then be compared to current productivity. Using these measures we find that, on average, at the time the survey was taken, agricultural production was about 68% of best production, with a standard deviation of 0.18. As a risk measure, we would expect that households with low yields relative to historical highs would be in a precarious situation with regard to savings, which would be depleted, and access to food, which would be scarce and rising in price. This, we surmise, would increase the demand for credit. On the other hand, the higher the yield risks, the less likely potential lenders would actually receive timely repayment. These two, not mutually exclusive, economic forces would likely result in more inelastic demand and supply relationships, meaning, as a matter of practice, that farmers in distress would willingly pay a higher interest rate in order to obtain much-needed liquidity and that creditors would charge a higher interest rate than would ordinarily be charged under better conditions. These better conditions would be captured by a rising value of the yield risk ratio to 1.0 (and higher if current production exceeds the previous historical high).

11.10 Hired Labor and Subsidiary Labor

Also critical to agricultural productivity is the use of labor. Labor can include household or family labor, and hired labor in terms of contributing to increased or improved productivity. During idle periods, excess household labor might be employed in subsidiary work or wage labor. Labor uses differed across regions. For example, Buck notes that 1/5th of all labor was performed by women, with the ratio being 31.6% in East Central China and 11.8% in North China—the latter percentage most likely due to the prevalence of foot binding in North China, particularly in the spring wheat area, where the binding is so tight as to compel women to undertake fieldwork on their knees.[2] Although we did have data on labor equivalents and hired labor, we took into consideration the possible multicollinearity between hired labor in units and farm size and other wealth/size covariates. Thus we include only a 1–0 binary variable to capture hired labor and subsidiary labor. The first captures the excess demand for labor which would be positive so long as the value of labor to agricultural productivity exceeded the prevailing wage rate, while the second captures an excess supply of labor with value in non-agricultural activities that provides greater value than labor on the farm would provide. Subsidiary work would most likely arise during periods of idleness, but could also be undertaken by less productive household members such as children, the elderly, or women with bound feet.

11.11 Time and Regional Variables

Buck's study took place in one of the more tumultuous periods in China's history. The period between 1929 and 1933 saw an almost relentless sequence of natural and man-made calamities, ranging from drought and famine to floods, insurgencies by communists, civil war and strife, Japanese annexation, and military action. It matters, then, when and where Buck collected data, since these events may have a bearing on the results. Of course, Buck would

[2] Buck (1937a) Op Cit. p. 292, Buck, J.L. (1930) p. 235.

not send his survey teams into war-torn areas, or areas held by communists, or even to drought- or flood-stricken areas, so for the most part we would expect that our investigations into credit supply and demand would be under near normal conditions. Nonetheless, these events, which might have affected the agricultural economies at the places and times of Buck's surveys, cannot be ignored and are included as instrumental variables (IV). To give a sense of the upheaval facing China's farmers over this period we provide a simplified summary in Table 11.7 which identifies the event (and, in the case of drought and flood, the year following an event) and provinces affected.[3]

We know from Buck's statistical summary when and where surveys were taken and to capture the effects we developed two sets of binary variables. We define nine time zones, identifying whenever the specific household data was recorded. They are: first half-year in 1929, second half-year in 1929, first half-year in 1930, second half-year in 1930, first half-year in 1931, second half-year in 1931, first half-year in 1932, second half-year in 1932, and year 1933. We also know the village/county, province, and agricultural region, and create corresponding binary variables for them as well. All told, our sample includes observations from 19 provinces, including Hebei, Shanxi, Liaoning, Jiangsu, Zhejiang, Anhui, Fujian, Shandong, Henan, Guangdong, Guangxi, Sichuan, Guizhou, Yunnan, Shaanxi, Gansu, Qinghai, Ningxia, and Suiyuan.

The specific events will not be captured by regional identifiers alone, for these will also capture the provincial politics and agricultural conditions generally present. However, the combination of the provinces and times not only allow us to control for fixed provincial effects but also the random effects of the calamities.

Buck said very little about these calamities. In both *Chinese Farm Economy* and *Land Utilization in China* the narrative is passive and neutral and never discussed in the context of what impacts such events might or might not have had on agricultural productivity and credit. To be fair, the survey part of *Land Utilization in China* as well as the randomization of localities to be surveyed was developed in 1928, just as the drought in

[3] The data provided in Table 10.7 are our own compilation from multiple sources, but largely from the reporting in *China Weekly Review*, a weekly English-language news magazine as reported separately in Chap. 3. All news items between mid-1928 and late 1933 were scanned for event information, including drought, famine, floods, bandits, communist activities, anti-bandit/anti-communist activities, warlord actions, anti-Japanese activities, and so on. Locations and dates were collated as best as we were able to the provinces and time schedule of the survey periods as reported by Buck.

Table 11.7 Summary of calamities facing Chinese farmers

	Major Drought	*Major Drought +1*	*Major Flood*	*Major Flood+1*	*Communism*	*Communist Suppression*	*Civil War*	*Northern War*	*Japanese Annexation of Manchuria*
Anhui	1	0	0	1	1	1	0	0	1
Fujian	0	0	0	0	1	0	0	0	1
Gansu	0	0	0	0	0	0	0	0	1
Guangdong	0	0	0	0	1	1	0	0	1
Guangxi	0	0	0	0	1	0	0	0	1
Guizhou	0	0	0	1	0	1	0	0	1
Hebei	0	0	0	0	0	0	0	0	1
Henan	0	1	1	1	1	1	0	1	1
Hunan	1	0	1	0	0	0	0	0	1
Hupeh	0	0	1	0	1	1	0	0	1
Jiangsu	0	0	1	1	1	0	0	1	1
Jiangxi	0	0	1	1	1	1	0	0	1
Liaoning	0	0	0	0	0	0	0	0	1
Ningxia	0	0	0	0	0	0	0	0	1
Qinghai	0	0	0	0	0	0	0	0	1
Shaanxi	1	1	1	0	0	0	1	0	1
Shandong	0	0	1	0	1	0	1	1	1
Shanxi	0	0	0	0	1	0	0	0	1
Sichuan	0	0	0	0	0	0	0	0	1
Suiyuan	1	0	1	0	0	0	0	0	0
Yunnan	0	0	0	1	0	1	0	0	1
Zhejiang	0	1	0	0	1	0	0	0	1

the north was beginning and before any of the other events occurred. Again, Buck avoided sending surveyors into distressed areas and even suspended the survey, setting his enumerators to work on flood relief in 1931. Nonetheless, these events should not be treated as production or credit neutral and our econometric approach sheds some light on the effects (see Chap. 3 and Dizi Chang 2014). A related problem is that we do not know within the six-month time bins we define whether a particular village was surveyed before, during, or after an event. A farm surveyed in the spring of one year would most likely refer to management practices and harvests for the previous year, while households surveyed at the end of a calendar year would likely respond in terms of the current year's harvest. To capture some of these effects we do also include a variable to capture households surveyed the year after a drought or flood. This would be meaningful for the 1928–1930 northern drought and the 1931 Yellow and Yangtze River floods, which were so severe that full recovery in the next crop year would have been unlikely for many farmers.

Although Buck did not correlate data with the events, he did query farmers on the extent of catastrophes, reporting that on average a Chinese farm household survived three famines in a lifetime, each lasting on average about 11 months, resulting in 5% of the population facing starvation and forcing 13% to emigrate. On top of this, each region on average faced 16 calamities, with the percentage of crop yield destroyed ranging between 43% and 54%.

11.12 Estimation Method

We have mentioned that the relationships between credit demand, credit supply, and agricultural productivity are endogenous. By endogenous, we really mean that they feed off each other. Our demand equation has an amount borrowed as a function of the interest rate on the loans, and under normal conditions we would expect that the higher the interest rate the lower the demand for credit. But the interest rate has to be agreed upon between the borrower and the lender and the lender might well set the rate depending, among other things, on the amount borrowed. A higher loan amount might come with a higher interest rate, but with this higher interest rate the borrower may reassess and reduce demand, which in turn will cause a revision in the interest rate offered and this will continue until a deal is struck. Likewise a farmer may have some savings with which to purchase seed or hire labor, but to obtain higher levels of economic efficiency it may be economically advantageous to borrow. Again, a dynamic arises between the borrower and lender in which a loan amount is requested and an interest rate offered, and if the interest rate extracts too much from the marginal value of production resulting from the loan then the borrower will revise accordingly and reduce the request and this will continue until a bargain is reached that returns a profit to both borrower and lender. In this way the demand for inputs drives a demand for credit. But if the bargain between borrower and lender succeeds in increasing output and profits, then as credit demand feeds production, production in turn feeds credit demand.

Our modeling approach is a three-equation simultaneously determined system of equations, with credit demand, credit supply, and agricultural productivity being treated as endogenous. To accommodate endogeneity we employ 3SLS, which is an appropriate instrumental variable technique when the error term is correlated with one or more explanatory variables. This is a reasonable assumption since we are assuming that the many extraordinary events between 1929 and 1933 are common sources of exogenous force that impact all three elements one way or another. From a statistical point of view the 3SLS estimator is biased in small samples but is consistent and asymptotically more efficient than single-equation estimators. Thus, it has desirable large sample properties.

The three equations to be estimated for credit demand, credit supply, and productivity respectively are:

$$\begin{aligned}\text{Loan Amount}_i &= \beta_0^1 + \beta_2^1\text{Interest Rate}_i + \beta_2^1\text{Productivity}_i + \beta_3^1\text{Wedding}_i \\ &+\beta_4^1\text{Dowry}_i + \beta_5^1\text{Birthday}_i + \beta_6^1\text{Birth of Son}_i + \beta_7^1\text{Funerals}_i \\ &+\beta_8^1\text{Other Special Expenditures}_i + \beta_9^1\text{Time}_i + \beta_{10}^1\text{Region}_i + \varepsilon_1\end{aligned}$$

$$\begin{aligned}\text{Interest Rate}_i &= \beta_0^2 + \beta_1^2\text{Loan Amount}_i + \beta_2^2\text{Productivity}_i + \beta_3^2\text{Yield Risk}_i \\ &+\beta_4^2\text{Relatives}_i + \beta_5^2\text{Friends}_i + \beta_6^2\text{Informal Business}_i + \beta_7^2\text{Semi} \\ &-\text{Formal Business}_i + \beta_8^2\text{All Other Borrowing Sources}_i \\ &+\beta_9^2\text{Time}_i + \beta_{10}^2\text{Region}_i + \varepsilon_2\end{aligned}$$

$$\begin{aligned}\text{Productivity}_i &= \beta_0^3 + \beta_1^3\text{Loan Amount}_i + \beta_2^3\text{Interest Rate}_i + \beta_3^3\text{Farm Area}_i \\ &+\beta_4^3\left(\text{Farm Area}\right)_i^2 + \beta_5^3\text{Labor Animals}_i + \beta_6^3\text{Productive Animals}_i \\ &+\beta_7^3\text{Income from Other Source}_i + \beta_8^3\text{Hired Labor}_i + \beta_9^3\text{Subsidiary Labor}_i \\ &+\beta_{10}^3\text{Time}_i + \beta_{11}^3\text{Region}_i + \varepsilon_3\end{aligned}$$

The credit demand equation has (with the exception of one model variant) the actual loan amount as the dependent variable and interest rate as the key endogenous independent variable. We include productivity to capture agricultural productivity and also include special expenditures. Much of the literature on credit in this era, anecdotal as it is, points to special expenditures as a major driver of farm debt, particularly for weddings and funerals. Wedding expenditures arise if a daughter is getting married. The expenditure will typically be paid by the girl's parents. But when a girl leaves a household, the family also loses a source of household labor and thus a dowry was often paid to the girl's household by the groom's family as a form of compensation. The time and region variations are random-effect binary variables for each of nine (six-month) periods between 1929 and 1933. The region variables are fixed-effect binary variables representing the provinces.

The second equation is the credit supply equation. This equation is interest-rate dependent, with the interest rate determined by the lender depending on a number of factors including the (endogenous) loan amount, the (endogenous) productivity of the borrower, and, to capture uncertainty, the yield risk variable. We also include the source of the loan. Here we want to examine whether the source of the loan matters. Typically we would expect that borrowing from friends and relatives would come at lower interest rates than formal and semi-formal businesses which could include usury rates. The random- and fixed-time and region effects are also included to account for exogenous factors that could affect credit supply. If a loan was made in a region facing drought or military incursion, it is possible that this would affect loan supply.

The third equation captures farm productivity effects and includes the (endogenous) credit effects of the loan amount (a source of liquidity with which to purchase inputs or labor) and the interest rate, which is a cost of doing business. We also include factors of production such as farm area (and its square to capture economies of size and scale), labor animals, and hired labor. Productive animals, subsidiary labor, and other income are also included to capture product substitution effects. The region and time effects again capture the location (approximate) in agricultural regions with the latter also capturing shorter term random effects.

We examine several variants of this model. The first model includes all respondents regardless of whether they were indebted or not. The dependent variable is equal to one if a loan was in place and zero otherwise. The interest rate in the supply equation is the weighted interest rate assigned to borrowers and zero otherwise. This is equivalent to a linear probability model (LPM). Normally, problems of this type would use a logit, probit, or tobit function at least for the demand equation, but, in a systems approach, blending a non-linear regression in a 3SLS with linear regressions can result in unstable standard errors. The LPM will generally provide the same story as the marginal effects of the logit/probit/tobit models but suffers from a predicted value that can (and most likely will) fall outside the {0,1} bounds of a probability distribution. Since we are interested in the explanatory power of the regression rather than a prediction, we can live with this sin. More problematic is the inclusion of interest rates and loan amounts in the first place. As will be seen, the coefficients of interest rate in the demand equations are positive and the coefficients of loan size in the supply equations are positive because, as a tautology, a zero loan is assigned a zero interest rate for non-borrowers while a positive rate is assigned a positive loan amount for borrowers. The coefficients in this first equation have no economic meaning and are included simply as control variables. Rather, what we are interested in with this binary demand structure is to get some idea of other factors that might affect whether a farmer borrows or not.

This is not the case for the remaining three models we present. The second model includes all loans, the third model considers only consumption loans, and the fourth model considers only production loans. These models include only borrowers and in these we seek to understand not only the economic relationships between loan amounts and interest rate to see whether demand is downward sloping and supply upward sloping as theory suggests, but also the household and economic drivers of agricultural credit demand and supply in China's Republican era.

11.13 Econometric Results: Who is Borrowing?

The results in Table 11.8 raise a simple question: Who are the borrowers and are there differences between the characteristics of borrowers or non-borrowers? These coefficients should be interpreted in the context of an increasing or decreasing likelihood of borrowing. Both ordinary least squares (OLS) ($N = 1718$) and 3SLS ($N = 1314$) regressions are provided. The greater reliance is on the 3SLS model. More detailed results follow, but the overall finding is that generally speaking there is little difference between the borrower and non-borrower subgroups. From the demand equation it does not appear that we can justify sweeping statements that special expenditures on weddings, funerals, birthdays, dowries, and so on

Table 11.8 Binary model of agricultural credit

Dependent variable (bold)	*OLS*		*3SLS*	
	coefficient	P *value*	*coefficient*	P *value*
Dummy loan				
Independent variable				
Weighted interest rate	0.0708975	0	0.2058877	0
Productivity	−0.0010522	0.317	0.0071722	0.103
Wedding	2.66E−07	0.993	−0.0000167	0.74
Dowry	−0.0000916	0.016	−0.0000516	0.645
Birthday	−0.0001437	0.003	−7.42E−05	0.737
Birth of son	−0.0004156	0.002	−0.0001806	0.618
Funeral	−4.00E−06	0.854	5.94E−07	0.987
Other	−0.0012864	0	−0.0005987	0.885
Year 1929 h1	0.1995966	0		
Year 1930 h1	0.1350179	0.259	−0.2972795	0.164
Year 1930 h2	0.1280955	0.007	0.1273513	0.059
Year 1931 h1	0.2065745	0.058	−0.3391054	0.172
Year 1932 h1	0.1927577	0.041	−0.3282276	0.141
Hebei	−0.1256535	0.168	0.1853788	0.228
Shanxi	0.032017	0.69	0.2838039	0.049
Guangxi	0.0326866	0.718	0.4342342	0.055
Guizhou	−0.0912546	0.194		
Yunnan	−0.0521477	0.513	0.324354	0.098
Gansu			−0.772441	0
Qinghai	0.1402073	0.143	0.4191007	0.044
Ningxia	0.1552007	0.051	0.3736906	0.013
Suiyuan				
Number of obs. = 1718			Number of obs. = 1314	
$F(19, 1698)$ = .			Parameters = 19	
Prob. > F = .			RMSE = 0.5620071	
R-squared = 0.4672			"R-sq." = −0.1529	
RMSE = 0.37116			chi^2 = 812.45	
			P = 0.0000	
Dependent variable				
Weighted interest rate				
Independent variable				
Dummy loan	0.6902606	0	0.9349819	0.763
Productivity	0.0162211	0.331	−0.0043298	0.87
Yield risk	−0.2709193	0.349	−0.071816	0.848
Relatives	1.510982	0	1.776068	0.352
Friends	3.442942	0	3.473832	0.195
Informal businesses	2.942236	0	2.53253	0.166
Formal/semi-formal businesses	1.775289	0	2.76759	0.355
All others	2.507147	0	2.698801	0.258
Year 1929 h1	−0.0217677	0.958		
Year 1930 h1	4.29044	0	0.2017773	0.869
Year 1930 h2	−0.1524913	0.252	−0.4218242	0.199
Year 1931 h1	3.695907	0.001	−0.2648421	0.874

(*continued*)

Table 11.8 (continued)

Dependent variable (bold)	*OLS*		*3SLS*	
	coefficient	P *value*	*coefficient*	P *value*
Year 1932 h1	4.258063	0	0.2931026	0.813
Hebei	−4.042747	0	0.3350961	0.764
Shanxi	−4.141461	0	0.1036986	0.929
Guangxi	−4.11885	0	−0.3016163	0.839
Guizhou	−4.560338	0		
Yunnan	−4.461356	0	−0.2643724	0.809
Gansu			4.430215	0
Qinghai	−3.861846	0	0.1658326	0.925
Ningxia	−4.931953	0	−0.7634071	0.424
Number of obs. = 1717			Number of obs. = 1314	
F(20, 1697) = 87.75			Parameters = 19	
Prob. > *F* = 0.0000			RMSE = 2.371333	
R-squared = 0.5625			"*R*-sq." = 0.5374	
RMSE = 2.0831			chi² = 1350.74	
			P = 0.0000	
Dependent variable				
Productivity				
Independent variable				
Dummy loan	−1.795239	0.253	9.688896	0.007
Weighted interest rate	0.5608175	0.339	−2.313831	0.003
Farm area square	−0.0002232	0	−0.0002252	0
Farm area local mou	0.0881802	0	0.0922535	0
Labor animal	0.3980929	0.096	0.4016467	0.119
Productive animal	0.0820473	0.638	−0.0145744	0.955
Other income %	403,455	0.129	0.0406555	0.498
Dummy hired labor	4.309383	0	3.3936	0
Dummy subsidiary	0.3550058	0.694	0.3100553	0.724
Year 1930 h1	−27.42663	0		
Year 1930 h2	−2.855484	0.037	−3.956419	0.018
Year 1931 h1	−6.763873	0.001	18.98447	0
Year 1932 h1	−11.66465	0	14.28114	0
Hebei	29.0319	0	2.976021	0.236
Shanxi	27.16659	0	0.729946	0.827
Guangxi			−26.44501	0
Guizhou	54.68648	0	30.75549	0
Yunnan	13.18848	0	−12.32866	0.002
Gansu	39.49775	0	27.37746	0
Qinghai	13.83657	0	−12.76711	0
Ningxia	26.86419	0	−0.5235277	0.888
Number of obs. = 1321			Number of obs. = 1314	
F(20, 1301) = 325.85			Parameters = 20	
Prob. > *F* = 0.0000			RMSE = 14.56467	
R-squared = 0.7755			"*R*-sq." = 0.7086	
RMSE = 12.869			chi² = 4390.91	
			P = 0.0000	

lead to credit use. We find that facing a special expenditure cannot be used as a matter of course to explain why a farm household borrows. Sometimes they do, sometimes they don't, with no statistical leaning one way or the other. We do find that, at a reasonable level of statistical reliability, higher productivity farmers are more likely to borrow. This paints a somewhat different story than the conventional wisdom might suggest. We cannot conclude that special expenditures are a driver of credit use, but can conclude that productivity is. When we consider the production equation in Table 11.8 we can observe the endogenous relationship between credit demand and productivity. Farms that borrow tend to be more productive, and more productive farms tend to borrow. We also find a size effect. Larger farms tend to be more productive and thus borrow more, while other factors that contribute to productivity such as hired labor and labor animals also then contribute to credit use.

11.14 Econometric Results: The Demand and Supply of Agricultural Credit

The next sets of econometric results explain what are best considered the conditional demand for credit. In Sect. 11.13 the assessment was based on all farmers for which data was available, including those who were not in debt. Here we now reduce the focus only to the farms who reported agricultural credit. First we examine any type of debt, and then look more closely at whether the debt was for productive or non-productive purposes. Consequently, the sample size reduces accordingly and the reader may want to keep this under consideration.

Table 11.9 shows the OLS and 3SLS results for total credit demand (productive and/or non-productive credit). The first regression is the demand equation, the second the supply equation, and the third the productivity equation. We focus again on the 3SLS results because they account for the endogeneity of demand, supply, and productivity. We find that more productive farms borrow more and that households with special expenditures for weddings and funerals also borrow more. We do not find that special expenditures on birthdays, the birth of a boy, or dowries lead a household into increased borrowing. The result is an interesting one because in the contemporaneous writings of the day there are many claims of farm households borrowing specifically for weddings and funerals. Our results support these claims.

Table 11.9 Total loan for indebted farmers with time and region variables

Dependent variable (bold)	*OLS*		*3SLS*	
	coefficient	P *value*	*coefficient*	P *value*
Total loan				
Independent variable				
Weighted interest rate	−0.5106107	0.338	14.53257	0.038
Productivity	0.1628833	0.611	2.888649	0.068
Wedding	6.68E−01	0	0.5694614	0
Dowry	0.0920703	0.597	0.0263822	0.861
Birthday	0.6717979	0.644	1.15E+00	0.847
Birth of son	0.4722442	0.448	0.7112581	0.263
Funeral	3.08E−01	0	2.01E−01	0
Year 1929 h1	2.913435	0.433		
Year 1930 h1	−38.25168	0.488		
Year 1930 h2	−6.153648	0.834	20.75009	0.373
Year 1931 h1	40.72046	0.296	37.78344	0.573
Year 1932 h1	5.642136	0.859	−6.392555	0.936
Hebei	24.04278	0.552	−75.53818	0.095
Shanxi	−37.04501	0.339	−124.3306	0.06
Guangxi	4.373326	0.921	−13.17376	0.872
Guizhou	14.31937	0.649	−104.297	0.179
Yunnan			212.2831	0.019
Gansu	0.6157157	0.984	−249.4231	0.02
Qinghai	−30.49923	0.478	−99.0438	0.133
Ningxia	−20.7167	0.552	−92.71529	0.194
Number of obs. = 439			Number of obs. = 361	
F(18, 420) = .			Parameters = 18	
Prob. > F = .			RMSE = 119.8012	
R-squared = 0.6904			"R-sq." = 0.5757	
RMSE = 95.9			chi^2 = 870.83	
			P = 0.0000	
Dependent variable				
Weighted interest rate				
Independent variable				
Total loan	0.0004069	0.539	0.0033668	0.166
Productivity	0.0754979	0.244	−0.078592	0.225
Yield risk	−2.674854	0.198	−2.533015	0.119
Relatives	0.155313	0.492	−0.2215084	0.687
Friends	−0.5745886	0.427	−0.2355776	0.702
Informal businesses	0.7314436	0.038	0.5368589	0.307
Formal/semi-formal businesses	0.4816642	0.191	0.2633986	0.731
All others	0.3265979	0.491	0.1510708	0.818
Year 1929 h1	3.906955	0.026		
Year 1930 h2	−0.7830909	0.112	−1.047519	0.294
Year 1931 h1	−3.856079	0.006	−0.3783078	0.891
Year 1932 h1	−3.822216	0.019	0.4274127	0.897
Hebei	4.534557	0.016	5.482501	0.004

(*continued*)

Table 11.9 (continued)

Dependent variable (bold)	*OLS*		*3SLS*	
	coefficient	P *value*	*coefficient*	P *value*
Shanxi	6.080447	0.002	5.841284	0.042
Guangxi	8.406599	0	3.964342	0.312
Guizhou	6.686425	0.001	6.89809	0.028
Yunnan	6.038755	0	2.130672	0.611
Gansu	14.06407	0	15.11412	0
Qinghai	7.139634	0	5.524437	0.051
Ningxia	6.922309	0.003	5.337925	0.103
Number of obs. = 440			Number of obs. = 361	
$F(20, 420) = 221.08$			Parameters = 19	
Prob. > $F = 0.0000$			RMSE = 4.342678	
R-squared = 0.6750			"R-sq." = 0.5738	
RMSE = 3.6091			$chi^2 = 632.47$	
			$P = 0.0000$	
Dependent variable				
Productivity				
Independent variable				
Total loan	0.0018664	0.463	0.0109645	0.388
Weighted interest rate	0.7389451	0.324	−3.872628	0.025
Farm area square	−0.000107	0.133	−0.0000661	0.501
Farm area local mou	0.0482158	0.217	0.0417356	0.36
Labor animal	0.423825	0.466	0.5651593	0.488
Productive animal	0.4415527	0.549	−0.5611113	0.531
Other income %	0.040442	0.66	0.0011485	0.995
Dummy hired labor	4.055596	0.001	2.872348	0.147
Dummy subsidiary	−1.763553	0.23	−1.189474	0.521
Year 1930 h2	−2.51142	0.286	−3.981249	0.376
Year 1931 h1	22.82909	0	16.08599	0.157
Year 1932 h1	25.29787	0	22.24962	0.094
Hebei	1.634978	0.663	17.02094	0.076
Shanxi	−4.073426	0.476	12.2727	0.364
Guangxi	−34.94398	0	−18.15344	0.211
Guizhou	0.5189212	0.947	22.11491	0.176
Yunnan	−24.52106	0	−15.76009	0.377
Gansu	1.184048	0.896	54.7489	0.022
Qinghai	−15.82887	0.002	5.272599	0.691
Ningxia	−10.32131	0.087	4.472979	0.759
Suiyuan				
Number of obs. = 364			Number of obs. = 361	
$F(20, 344) = 86.68$			Parameters = 20	
Prob. > $F = 0.0000$			RMSE = 21.8244	
R-squared = 0.8013			"R-sq." = 0.3049	
RMSE = 12.016			$chi^2 = 531.43$	
			$P = 0.0000$	

Another point of note is that there is a positive relationship between the amount borrowed and the interest rates charged on loans. The interest rate charged on loans is determined by a number of factors including loan amount ($p = 0.166$) and yield risk ($p = 0.11$). On this latter point a higher number means that current productivity is getting closer to historical best yields so it does appear that the lender takes risk into consideration, charging a higher interest rate for higher risk farms. However, with the 3SLS results we find no evidence that one source of credit charges an interest rate higher than any other when other things are considered such as time and province. However, the OLS results do suggest that informal businesses ($p = 0.038$) do charge a higher rate, although we place less weight on this result.

The loan–interest rate results are interesting. Taking the positive loan amount to interest rate from equation 1 and the positive interest rate to loan amount in equation 2 suggests that farmers are interest-rate takers. In other words, this result can only be consistent if the actual demand facing the farmer is highly inelastic so that a higher interest rate does not affect the borrowing decision. This is especially true for weddings and funerals where failure to provide an appropriate feast is regarded as humiliating. On the lender's side the decision on what interest rate to charge is quantity dependent; the higher the loan, the higher the interest rate charged with some adjustment for risk and some differentiation by use.

11.15 Econometric Results: The Demand and Supply of Non-Productive or Consumption Credit

Results for consumption loans are reported in Table 11.10. Again we find with strong significance that weddings and funerals are key drivers of non-productive loans. But we also find that larger consumption loans are positively related to productivity. This holds when productivity and credit are treated endogenously but does not hold with the OLS regression. We also find, again, that the demand curve is upward sloping and significant in the 3SLS model, but not significant in the OLS one. These we see as the same thing and conclude as before that the demand is almost perfectly, if not highly, inelastic so that farmers willingly accept a higher rate on larger loans. To the extent that there is negotiation between borrower and lender the story weakens. We find in the supply equation no statistical

Table 11.10 Consumption credit

Dependent variable (bold)	*OLS*		*3SLS*	
	coefficient	P *value*	*coefficient*	P *value*
Consumptive loan				
Independent variable				
Consumptive interest rate	−0.2344526	0.63	19.12231	0.002
Productivity	0.1657111	0.56	3.771076	0.026
Wedding	5.79E−01	0	0.4581681	0
Dowry	0.0889123	0.635	0.0130958	0.93
Birthday	0.9983116	0.394	1.20E+00	0.822
Birth of son	−1.24867	0.326	−0.5505292	0.64
Funeral	3.04E−01	0	1.94E−01	0
Other				
Year 1929 h1	2.425001	0.455		
Year 1930 h1	17.487	0.712	−258.2811	0.018
Year 1930 h2	15.25516	0.62	40.89557	0.117
Year 1931 h1	65.32261	0.001	−276.7796	0.005
Year 1932 h1	21.34468	0.055	−336.8034	0.001
Hebei	−28.27701	0.295	179.4723	0.031
Shanxi	−57.98391	0.001	170.5514	0.007
Guangxi	−6.34989	0.743	331.1812	0.001
Guizhou			207.7009	0.001
Yunnan	1.482697	0.969	551.327	0
Gansu	−16.77768	0.118		
Qinghai	−52.23498	0.028	195.6869	0.006
Ningxia	−30.3705	0.043	211.5881	0.002
Suiyuan				
Number of obs. = 381			Number of obs. = 315	
F(18, 362) = .			Parameters = 18	
Prob. > *F* = .			RMSE = 133.5065	
R-squared = 0.7089			"*R*-sq." = 0.3950	
RMSE = 87.491			chi^2 = 711.89	
			P = 0.0000	
Dependent variable				
Consumptive interest rate				
Independent variable				
Consumptive loan	0.0005842	0.485	0.004384	0.196
Productivity	0.0878831	0.21	−0.109683	0.147
Yield risk	−3.185998	0.169	−2.171547	0.234
Relatives	0.2931582	0.277	−0.2079623	0.711
Friends	−0.5190018	0.496	−0.1950251	0.753
Informal businesses	0.9270542	0.031	0.4542665	0.42
Formal/semi-formal businesses	0.7886978	0.142	0.252894	0.771
All others	0.6294758	0.161	0.2862689	0.678
Year 1929 h1	3.295705	0.082		
Year 1930 h2	−0.8595934	0.133	−0.9440116	0.447
Year 1931 h1	−4.119966	0.013	0.9192048	0.81

(*continued*)

Table 11.10 (continued)

Dependent variable (bold)	*OLS*		*3SLS*	
	coefficient	P *value*	*coefficient*	P *value*
Year 1932 h1	−3.885705	0.031	2.025213	0.64
Hebei	5.074939	0.013	5.177187	0.035
Shanxi	6.200795	0.005	4.799157	0.217
Guangxi	7.666323	0.004	1.045889	0.835
Guizhou	6.399851	0.003	5.096017	0.218
Yunnan	6.113548	0.001	−0.042744	0.994
Gansu	13.93146	0	14.3845	0
Qinghai	7.331143	0	4.313716	0.264
Ningxia	7.367471	0.003	4.346178	0.312
Number of obs. = 382			Number of obs. = 315	
$F(20, 362)$ = 126.00			Parameters = 19	
Prob. > F = 0.0000			RMSE = 4.812568	
R-squared = 0.6691			"R-sq." = 0.5255	
RMSE = 3.827			chi^2 = 518.41	
			P = 0.0000	
Dependent variable				
Productivity				
Independent variable				
Consumptive loan	0.0037018	0.243	0.0178198	0.301
Consumptive interest rate	0.7777542	0.302	−4.059061	0.018
Farm area square	−0.0000867	0.272	−0.0000506	0.635
Farm area local mou	0.0383796	0.376	0.028991	0.57
Labor animal	0.3087556	0.613	0.4524751	0.585
Productive animal	0.6321546	0.42	−0.400096	0.674
Other income %	0.1076263	0.396	−0.0030882	0.99
Dummy hired labor	3.801197	0.005	2.061805	0.316
Dummy subsidiary	−1.978248	0.212	−0.8249456	0.679
Year 1930 h1	−29.38809	0	52.89896	0.04
Year 1930 h2	−0.8802852	0.743	−2.895373	0.604
Year 1931 h1	−3.543324	0.422	72.17854	0
Year 1932 h1	−0.3640374	0.917	79.37668	0
Hebei	27.61863	0	−37.5833	0.053
Shanxi	21.78675	0	−44.35667	0.004
Guangxi	−7.475689	0.071	−77.4175	0
Guizhou	23.08524	0	−40.05429	0.004
Yunnan			−76.22105	0
Gansu	26.99077	0		
Qinghai	10.74549	0.011	−49.96959	0.001
Ningxia	14.93683	0	−50.81258	0
Number of obs. = 318			Number of obs. = 315	
$F(20, 298)$ = 80.64			Parameters = 20	
Prob. > F = 0.0000			RMSE = 23.74518	
R-squared = 0.7971			"R-sq." = 0.1751	
RMSE = 12.181			chi^2 = 413.88	
			P = 0.0000	

relationship between the weighted average rate of interest charged and the loan amount (p = 0.19) and no significance in the OLS regression either. Nor do we find price discrimination amongst the various suppliers, with none of them in the 3SLS or OLS models charging interest rates that statistically differ from the average.

When isolating consumption loans in this way it comes to mind that when a loan request for a wedding or a funeral is made, there is an inherent rate charged by all suppliers that is simply accepted by the borrowers. Unfortunately, this explanation is inadequate since it does a poor job of explaining the demand–supply paradox for consumption loans. Recall that the same evaluation for total indebtedness supports the view of demand inelasticity without paradox. Another explanation, which we can only conjecture for we have not seen this discussed in the contemporaneous literature of the time, is that there was in fact an excess supply of credit and farmers actually had bargaining power. The bargaining power could arise from social pressures and peer effects as well as reciprocity in the sense that today's lender may be tomorrow's borrower. Social pressures would frown upon usury and gouging and this might place a cap on the interest rates charged. If a farmer could approach multiple lenders with excess credit they may have to compete at rates below the social maximum. Meanwhile, a farmer would not necessarily balk at the idea of paying a higher interest rate for a higher loan, for when the day comes that they become a lender (or had in the past) they too would likely require the same.

Also interesting is the finding that neither productivity nor yield risk has a statistical impact (although both come with the expected negative sign), which suggests that at least from the supply side lenders do not put as much weight on these factors, likely because the special expenditures are unrelated to production and are deemed to be a social requirement of village life. In the productivity equation we find that there is no relationship between special expenditures and productivity, which gives credence to the argument that they are not substitutes for each other. This interesting results bolsters the argument that when regional differences and time frames are accounted for there is separability between consumption and production, at least when it comes to agricultural credit. It may be true that farmers have to give up other consumption items such as health or education or even food to repay the loans, but they do not sacrifice production along the way.

This does not hold true for the interest rates charged, however. There is a negative relationship between interest rates and productivity and this is significant in the 3SLS equation ($p = 0.018$) but not the OLS ($p = 0.302$).

From an economic point of view farmers may consider loan principal, a source of liquidity, as being fungible. The rate of interest charged, however, is a necessary expense of celebrating the wedding or funeral, and the higher this rate the more the farmer has to give up in terms of inputs to the production process.

11.16 Results Analysis: Production Loan for Indebted Farmers with Time and Region Variables

The final investigation is on production loans, of which there are only 58 observations for the unrestricted model, barely enough to be conclusive. Results are shown in Table 11.11. Here we find a negative relationship between interest rates and loan amounts, as would be expected in conventional credit theory. This holds in both the 3SLS ($p = 0.053$) and OLS ($p = 0.099$) models. We also find a positive relationship between productivity and loan amount ($p = 0.003$), suggesting that, as previously found, more productive farms tend to borrow more for production purposes. This is not symmetric though, since, unlike consumption loans, we do not find in the productivity equation that higher interest rates lower productivity. This suggests that as far as production is concerned the value of the input investments at least equals, if not exceeds, the cost of borrowing. (We also find, but with no specific interpretation, that there is a positive relationship between wedding ($p = 0.011$) and birth-of-son ($p = 0.057$) and production loans. This may be evidence of fungibility between production loans and special expenditures, but the granularity of the data does not support such a conclusion with any degree of comfort. It is safer to consider these to be statistical artifacts.)

On the supply side we find the relationship between loan amount and interest rates to be negative ($p = 0.053$). This contrasts with previous results that show a positive relationship between the loan amount and interest rates. The economic explanation for this result is that when it comes to production loans farmers have considerable market power in negotiating favorable terms for higher loan amounts. It is possible that lenders with excess cash would prefer to make a smaller number of higher valued loans to reduce transactions, regularizing, and collection costs. This argument is weakened by the finding that higher productivity farms pay a higher interest rate ($p = 0.064$), even though there is no statistical support in the productivity regression for any relationship between interest rates and productivity ($p = 0.409$), let alone the loan amount and productivity ($p = 0.0859$).

Table 11.11 Productive loan for indebted farmers with time and region variables

Dependent variable (bold)	*OLS*		*3SLS*	
	coefficient	P *value*	*coefficient*	P *value*
Productive loan				
Independent variable				
Productive interest rate	−13.81784	0.099	−55.33578	0.053
Productivity	0.4301687	0.694	5.440612	0.003
Wedding	2.23E−01	0	0.130652	0.011
Dowry	0.0906299	0.639	−0.0036724	0.988
Birthday	0		0.00E+00	
Birth of son	1.175789	0	1.049963	0.057
Funeral	2.90E−01	0.002	1.59E−01	0.224
Other				
Year 1929 h1	30.2709	0.218		
Year 1930 h1	−55.07195	0.668		
Year 1930 h2	0.7134642	0.985	46.6263	0.338
Year 1931 h1	11.21739	0.916	−79.99474	0.371
Year 1931 h2				
Year 1932 h1	35.43054	0.525	−47.79411	0.68
Hebei	75.14033	0.423	81.36491	0.446
Shanxi	58.75692	0.456	106.869	0.48
Guangxi	46.10739	0.54	246.2551	0.116
Guizhou	−7.706453	0.88	−132.4513	0.339
Yunnan	−9.285127	0.785	239.4167	0.105
Gansu			103.3964	0.651
Qinghai	38.42489	0.673	244.8395	0.159
Ningxia	−20.70702	0.548	32.9698	0.807
Number of obs. = 73			Number of obs. = 58	
$F(17, 55)$ = .			Parameters = 17	
Prob. > F = .			RMSE = 85.8562	
R-squared = 0.7475			"R-sq." = 0.5811	
RMSE = 69.325			chi^2 = 163.24	
			P = 0.0000	
Dependent variable				
Productive interest rate				
Independent variable				
Productive loan	−0.0017912	0.086	−0.0034321	0.053
Productivity	0.0108236	0.332	0.0387461	0.064
Yield risk	−0.5487945	0.576	0.4501207	0.644
Relatives	−0.1924474	0.595	−0.3930662	0.512
Friends	0.5742167	0.168	0.6975957	0.129
Informal businesses	0.1524437	0.51	0.0507588	0.868
Formal/semi-formal businesses	0.1702209	0.532	−0.0465682	0.923
All others	0.1543696	0.595	−0.0490804	0.929
Year 1929 h1	2.50171	0.004		
Year 1930 h2	−1.062789	0.001	−0.3346021	0.505
Year 1931 h1	−2.497807	0	−2.336454	0.003
Year 1932 h1	−2.872495	0.003	−2.511676	0.05

(*continued*)

Table 11.11 (continued)

Dependent variable (bold)	*OLS*		*3SLS*	
	coefficient	P *value*	*coefficient*	P *value*
Hebei	4.081876	0	2.623176	0.007
Shanxi	5.671531	0	3.984983	0.002
Guangxi	5.433001	0	4.598918	0.003
Guizhou	3.29561	0.014	1.30846	0.432
Yunnan	5.607441	0	4.121536	0.009
Gansu	7.268536	0	5.078707	0.001
Qinghai	6.216929	0	5.285053	0
Ningxia	4.133177	0.002	2.494583	0.078
Number of obs. = 73			Number of obs. = 58	
F(19, 53) = .			Parameters = 19	
Prob. > F = .			RMSE = 0.7673681	
R-squared = 0.9146			"R-sq." = 0.8891	
RMSE = 0.83746			chi^2 = 605.81	
			P = 0.0000	
Dependent variable				
Productivity				
Independent variable				
Productivity loan	−0.0285909	0.236	0.0057884	0.859
Productive interest rate	0.3019306	0.841	3.575461	0.409
Farm area square	−0.0002404	0.052	−0.0001679	0.204
Farm area local mou	0.1741037	0.042	0.1299531	0.169
Labor animal	1.332393	0.572	0.70695	0.733
Productive animal	−1.57288	0.337	−0.6992099	0.67
Other income %	−0.1469099	0.352	−0.1079454	0.568
Dummy hired labor	5.324517	0.057	4.310013	0.096
Dummy subsidiary	−4.55999	0.273	−4.224089	0.315
Year 1930 h1	32.72912	0.124		
Year 1930 h2	−10.97899	0.153	−9.972679	0.059
Year 1931 h1	40.85605	0.003	13.48962	0.221
Year 1932 h1	34.21577	0.003	7.076156	0.638
Hebei	−16.65698	0.181	4.754758	0.788
Shanxi	−13.20614	0.024	5.403447	0.824
Guangxi	−63.21222	0	−39.36868	0.079
Guizhou	17.44856	0.018	41.34098	0.041
Yunnan	−24.23681	0.026	−10.37198	0.669
Gansu			12.36935	0.708
Qinghai	−39.7622	0	−23.05063	0.341
Ningxia	−21.55644	0.003	2.239625	0.907
Suiyuan				
Number of obs. = 58			Number of obs. = 58	
F(19, 38) = .			Parameters = 20	
Prob. > F = .			RMSE = 8.616028	
R-squared = 0.8987			"R-sq." = 0.8829	
RMSE = 9.8973			chi^2 = 515.13	
			P = 0.0000	

11.17 Discussion on Credit Demand and Supply

There is much anecdotal evidence about credit demand and supply in the lead up to the formalization of credit cooperatives and formal agricultural banking during the Republican era, but what is missing is a thorough examinations of the economic characteristics of credit supply and demand. The conventional wisdom was that farmers en masse had an excess demand for credit and were forced in desperation to borrow from usurers of one sort or another at exorbitant rates. Our findings dispel some of these notions.

The discovery of John Lossing Buck's credit data in 2000 provides the first opportunity to assess credit demand and supply. Over 4000 records were recovered and of these over 3000 had actual (useable) data of agricultural credit, including production and non-production loan amounts, interest rates, productivity, risk, and special expenditures. From these we constructed four models: the first to investigate the characteristics of those who borrowed versus those who did not, and then from the subsample of those who borrowed the demand and supply relationships between total borrowing, borrowing for consumption (special expenditures), and borrowing for production. Although we present OLS results, we focus primarily on 3SLS results on the premise that supply, demand, and productivity are endogenously related.

Unfortunately, not all data from Buck was retrieved and for some variables we thought important there was not consistency in collection. Data such as farm size, for example, was not necessarily collected for all farms that had credit, and credit was not collected for all farms for which farm size was collected. This involved some sacrifices for statistical efficiency. When examining all loans we had only 361, 315, and 52 farms for all loans, consumption loans, and production loans respectively. This we understand is inadequate and we do our best not to oversell the results, and indeed provide the actual summary tables published by Buck in *Land Utilization in China* to provide requisite balance.

Nonetheless, there are some elements of credit demand and supply that we feel are robust and important. First, against the narrative that special expenditures lead to heightened credit use, we find no evidence that this is either necessary or sufficient. When using special expenditures in the demand equation, none come up significantly to explain which farmers borrow versus who do not. Observing a farmer has a loan does not imply special expenditures, nor does an observation of a special expenditure imply that farmers are driven into debt. Of course, we are being very general for we do not observe farms that wanted credit for a special expenditure and did not receive it.

A second result that we believe to be important we draw from the more reliable second regression on total indebtedness. There we find a very interesting dynamic between supply and demand. We believe that the demand for credit is nearly, if not perfectly, inelastic, meaning that farmers are price takers at any loan amount. The lender exploits this inelasticity by increasing the interest rate as the loan demand increases, as a local monopolist might do. We find the same for consumption loans and offer some strong statistical evidence that the reasons that farmers borrow for non-productive use is in part to cover special expenditures on weddings and funerals. We mentioned that this type of expenditure was very much a part of Chinese culture and status seeking, as well as about saving face. These results are very much in line with some of the discussions in the Chinese literature at the time.

Results for production loans are far less reliable, but interestingly show some degree of demand elasticity and negative slope. This is an appealing result, with the direction of causality going from higher interest rates to lower amounts borrowed. But on the supply side we also find a negative relationship in the causal direction of a large loan leading to a lower interest rate. If the observed interest rates were an autarky result we would expect that the supply result would be positive as found for consumption and total indebtedness. The results for production credit are unconvincing one way or another since, by construction of the 3SLS structure, failure to explain one result means that all results are unreliable, even if at first glance they make economic sense.

Acknowledgements This chapter is modified from H. Fu and C.G. Turvey (2018). "Estimating the Demand for Farm Credit in the Republican Era." In *The Evolution of Agricultural Credit during China's Republican Era, 1912–1949* (pp. 179–232). Palgrave Macmillan, Cham. We thank Dizi Chang (2014) and Ziang Cheng (2016) for assisting in the identification of conflicts and calamities and the econometric work presented here.

References

Buck, J.L. 1930. *Chinese Farm Economy: A Study of 2866 Farms in Seventeen Localities and Seven Provinces in China*. Chicago: University of Chicago Press.

———. 1937a. *Land Utilization in China: A Study of 16,786 farms I 168 localities, and 38,256 farm families in twenty-two provinces in China, 1929–1933*. University of Nanking (1937), University of Chicago Press, Chicago (1937), and The Council on Economic and Cultural Affairs, NY NT, 1956.

Chang, Dizi. 2014. *An Economic Analysis of Calamities and Conflicts in Rural China: 1929–1933*. Unpublished MS Thesis, Dyson School of Applied Economics and Management, Cornell University.

Cheng, Ziang. 2016. *Agricultural Credit Demand in Republican China, 1929–1933: An Econometric Assessment of John L. Buck's Household Survey Data*. Unpublished MS Thesis, Dyson School of Applied Economics and Management, Cornell University.

Fu, H., and C.G. Turvey. 2018. Estimating the Demand for Farm Credit in the Republican Era. In *The Evolution of Agricultural Credit during China's Republican Era, 1912–1949*, 179–232. Cham: Palgrave Macmillan.

Kumar, C.S., C.G. Turvey, and J.D. Kropp. 2013. The Impact of Credit Constraints on Farm Households: Survey Results from India and China. *Applied Economic Perspectives and Policy* 35 (3): 508–527.

Turvey, C.G., G. He, M.A. Jiujie, R. Kong, and P. Meagher. 2012. Farm Credit and Credit Demand Elasticities in Shaanxi and Gansu. *China Economic Review* 23 (4): 1020–1035.

Turvey, C.G., and R. Kong. 2010. Informal Lending Amongst Friends and Relatives: Can Microcredit Compete in Rural China? *China Economic Review* 21 (4): 544–556.

Turvey, C.G., R. Kong, and X. Huo. 2010. Borrowing Amongst Friends: The Economics of Informal Credit in Rural China. *China Agricultural Economic Review* 2 (2): 133–147.

Verteramo-Chiu, L., V. Khantachavana S., and G. Turvey C. 2014. Risk Rationing and the Demand for Agricultural Credit: A Comparative Investigation of Mexico and China. *Agricultural Finance Review* 74 (2): 248–270.

CHAPTER 12

The Change in China's Cropland Utilization and Productivity Over Nearly a Century in China: A Comparison Study Based on Buck's Survey

Hao Hu and Feng Zhang

12.1 Introduction

In this chapter we analyze the cropland utilization of China over the past century by comparing relevant data from Buck and Buck's survey data to cropland utilization in present-day China (Buck 1937a, b, c). During different historical periods, farm households faced different economic and policy conditions and thus had significantly different land use practices. As a limited natural resource, the demand for land increases with economic development and population. China's unprecedented growth since 1978

H. Hu (✉)
College of Economics and Management, Nanjing Agricultural University, Nanjing, China
e-mail: huhao@njau.edu.cn

F. Zhang
Jiangsu Academy of Agricultural Sciences, Nanjing, China

H. Hu et al. (eds.), *Chinese Agriculture in the 1930s*,
https://doi.org/10.1007/978-3-030-12688-9_12

is a case in point. Not only has industrial growth and urbanization placed pressures on agricultural land, but modern developments in agricultural technologies have shifted not only the input mix (e.g. use of chemical fertilizers, etc.) but also productivity and the crop mix. Most studies examining changes in land utilization over the past century or so have relied heavily on aggregated macro data, and this includes prior references to the aggregated form of Buck's data in his book and statistical summary. In this chapter we take advantage of the micro data now available to us to investigate these differences and in so doing provide additional insights into the historical development and *Land Utilization in China*.

12.2 Selection of Related Indexes and Areas for Comparison.

Before conducting our comparative analysis, we shall define the comparison objects and concepts and select the specific comparison indexes. Cropland utilization refers to the various operating activities on cropland performed by farm households, including agricultural and non-agricultural activities. During the 1920s–1930s farmers were firmly bound to the land. Stuck with traditional production practices, a traditional social system, and an almost feudal land system there were no (or limited) non-agricultural operation activities of consequence. Focusing on farm households' agricultural operation activities we use and develop indexes for measurements of cropland utilization that include per household cropland area, degree of land fragmentation, cropping structure, multiple cropping index (MCI), cropland degree of intensity, and cropland productivity. Cropland productivity is an important index for evaluating land utilization performance. It captures the output or value of the output per unit area of cropland or yield of various production factors input. Indexes in this study include the cropland productivity and agro-labor productivity of farm households.

In his survey of China's land utilization status, Buck divided the survey sample regions into two major agricultural zones—wheat zone and rice zone—and these were further divided into eight areas, namely, Spring Wheat Area, Winter Wheat–Millet Area, Winter Wheat–Kaoliang Area, Yangtze Rice–Wheat Area, Rice–Tea Area, Sichuan Rice Area, Double-Cropping Rice Area, and Southwestern Rice Area (see also Fig. 2.1 and 2.2 in Chap. 2). The division of agricultural zones and areas was based on the natural endowment of agricultural production and the dominance of particular crops cultivated. The survey areas covered agricultural areas of varying weather conditions. For comparison purposes, we select one typical province for each region. Specifically, Gansu Province is selected as a typical province for the Spring Wheat Area, Shaanxi Province for the Winter Wheat–Millet Area, Shandong Province for the Winter Wheat–Kaoliang Area, Jiangsu Province for the Yangtze Rice–Wheat Area, Zhejiang Province for the Rice–Tea Area, Sichuan Province for the Sichuan Rice Area, Guangdong Province for the Double-Cropping Rice Area, and Guizhou Province for the Southwestern Rice Area.

12.3 Comparison of the Degree of Land Utilization

12.3.1 *Change in Per Household Cropland Area*

Over the twentieth and twenty-first centuries, the per household cropland area has decreased sharply. In 2005 it was less than one third of that in the 1920s–1930s. This change was not due to natural economic conditions, but rather the era of collectivization between 1949 and 1978, and the era of reforms under the household responsibility system and issuance of land use rights (LUR) between 1978 and the present day. Comparing data of the two major agricultural zones, we find that the per household cropland area in the wheat zone was greater than that in the rice zone in both periods, indicating the inconsistent distribution of cropland resources and hydrothermal resources (Table 12.1). As compared with that of the 1920s,

Table 12.1 Comparison of per household cropland area between 1929–1933 and 2005 (hectare)

Agricultural area	*1929–1933*	*Region*	*2005*
Eight-area average	1.51	National average	0.50
Spring wheat	2.96	Gansu	1.08
Winter wheat–millet	1.50	Shaanxi	0.73
Winter wheat–sorghum	2.06	Shangdong	0.38
Yangtze rice–wheat	1.42	Jiangsu	0.32
Rice–tea	0.88	Zhejiang	0.17
Sichuan rice	1.27	Sichuan	0.46
Double-cropping rice	0.88	Guangdong	0.21
Southwestern rice	0.81	Guizhou	0.62

Source: Data for 1929–1933 are from *Land Utilization in China* (John L. Buck), p. 352. Data for 2005 are from *China Rural Statistical Yearbook*, calculated and rearranged by the authors

Note: Due to lack of statistical data, the per household cropland areas for 2005 were calculated by dividing the total area of cropland of each province with the total number of farm households in the province.

the per household cropland area in the wheat zone in 2005 fell by 66%. The per household cropland area had a sharper decline in the rice zone, particularly in economically better developed provinces such as Jiangsu, Zhejiang, and Guangdong. Notwithstanding the massive disruption to land use, ownership, practices, and tenancy during the 1950–1978 era of collectivization, the past decades have witnessed a drastic increase in China's total and rural population, which has led to decreases of per household cropland area. Additionally, China's industrial advance and urban development have occupied a great deal of agricultural land and caused a decline in cropland area. As a result of economic development and continuous growth of population, the conflict between population and land resources has become more and more prominent.

12.3.2 *Change in Land Fragmentation per Household*

Land fragmentation is one of the most important characteristics of China's agriculture. Generally, it is measurable by indexes such as the number of cropland parcels owned by farm household, average size of land parcels, and average distance from land parcels to farmstead. The degree of cropland fragmentation is affected by factors such as topography, land system, and technical level. Objectively, land fragmentation has caused disadvantages to agricultural production, including limitation on the use of agricultural machinery, increased farm household transportation cost, irrigation cost, and farming management difficulty. These factors have weakened the comparative advantage of agriculture, and it has only been recently that land reforms have been promulgated to address these issues.

Observed from data presented in Table 12.2, cropland fragmentation in 1929–1933 was very general, with the average number of cropland parcels per farm households as high as 5.6 and the average size of cropland parcel as little as 0.38 hectare. Notably, the number of parcels in the Sichuan Rice Area was 9.7, with an average size of 0.55 hectares and a distance of about 300 meters between plots. In the other rice areas there were fewer plots, of smaller size, but more distance between them. These distances and the number of plots could be related to the tenancy of the land and subdivision by landlords, but could also be attributed to the terrain required for growing rice and the development of paddy fields. In terms of dry grains the Spring Wheat Area averaged 4.8 parcels per farm, each of about 0.92 hectares on average, but the fragmentation was largely by distance. Fields were on average 1 km apart.

Table 12.2 Situation of land fragmentation per household in 1929–1933

Agricultural area	*Average number of parcels per household*	*Average size of parcel (hectare)*	*Average distance between parcels and farmsteads (km)*
Average	5.6	0.38	0.6
Spring wheat	4.8	0.92	1.0
Winter wheat–millet	5.5	0.30	0.9
Winter wheat–kaoliang	6.2	0.40	0.6
Yangtze rice–wheat	5.3	0.41	0.3
Rice–tea	5.4	0.18	0.6
Sichuan rice	9.7	0.55	0.3
Double-cropping rice	5.4	0.23	0.7
Southwestern rice	3.7	0.26	0.6

Source: ***Land Utilization in China***, pp. 216–224

Table 12.3 Comparison of land fragmentation per household between 1929–1933 and 2006

Agricultural area	*Average number of parcels per household*	*Average size of parcel (hectare)*	*Average distance between parcels and farmstead (km)*
Yangtze Rice–Wheat Area (1929–1933)	5.3	0.4	0.3
Jiangsu Province (2006)	4.7	0.08	0.5

Source: Data for 1929–1933 are from ***Land Utilization in China***, p. 220. Data for 2006 are from ***Cropland Fragmentation, Labor Utilization and Peasants Income*** (Li Gongkui), Nanjing Agricultural University Ph.D. thesis, 2006.

Due to limitations in the availability of contemporary data, we can only make a simple comparison between the 2006 cropland fragmentation survey data for the northern districts of Jiangsu Province with the 1929–1933 survey data for the Yangtze Rice–Wheat Area. The result of the comparison is presented in Table 12.3.

Comparison of data for these two periods reveals that the problem of cropland fragmentation has not been eased after nearly a century. The average size in Jiangsu of 0.08 hectares is fairly close to the size of one modern mou (1/15th of a hectare or about 0.067 hectares) so it is likely that part of the fragmentation is directly related to the allocation of LUR. The number of cropland parcels owned by farm households in these two periods are roughly equivalent, and the absolute increase of population reduced the per household cropland areas and dramatically reduced the size of each parcel. Again, in the modern era the allocation of non-contiguous LUR for each family member probably plays a significant role in understanding these numbers. In comparison, the number of parcels in 1929–1933 Jiangsu would more than likely result from tenancy relationships where, over time, farm households would enter into rental arrangements as land parcels for rent became available. Needless to say, the degree of land fragmentation and the small parcel size would have greatly restricted the application of machinery and other technology investments in China's agriculture (and in many cases still does).

12.3.3 Changes in Cropping Structure

An investigation of farm households' cropping structures may reflect households' land utilization preferences in these two periods and the change in China's cropping structure.

The determinant factors of China's agricultural cropping structure in 1929–1933 were just the endowment of natural resources like soil and climate. Due to a lower level of productivity and unstable social and political environment, the crops planted in that period were mainly rice, wheat, kaoliang, potato, and cotton, which provided food and clothing. Cash crops planted in some regions, for example opium planted in the Southwestern Rice Area and Sichuan Rice Area, were owned by warlords and prominent landlords.

Since China's reform and opening up in the early 1980s, agricultural productive forces have been greatly liberated by the reform of agricultural policy and systems, especially the establishment of the household contract responsibilities system. The development of a post-collective market economy as well as improvements in urban and rural living standards have influenced the modern cropping structure. Table 12.4 shows a decline in the proportions of land planted with staple grain crops such as rice and wheat, as compared with that of 1930s. The planting areas of rice and wheat fell by 15% and 14%, respectively, while the planting areas of corn, soybean, and oil crops substantially increased.

Table 12.4 Comparison of crops' planting area proportion between 1929–1933 and 2005 (%)

Agricultural area	*1929–1933*	*Region*	*2005*
Average of eight areas	Rice 33; wheat 29; cotton 7	National average	Rice 18; wheat 14; cotton 4; corn 17; bean and pea 8; tuber crop 6; oil 9; vegetable 15; fruit 10
Spring wheat	Millet 34; spring wheat 18; potato10	Gansu	Wheat 27; cotton 2; corn 13; bean and pea 6; tuber crop15; oil 9; vegetable 9; fruit 10
Winter wheat–millet	Wheat 40; millet 21; cotton 9	Shaanxi	Rice 3; wheat 28; cotton 2; corn 26; bean and pea 9; tuber crop 8; oil 6; vegetable 8; fruit 20
Winter wheat–kaoliang	Wheat 46; millet 23; cotton 9; kaoliang 19; corn 16	Shandong	Wheat 31; cotton 9; corn 26; bean and pea 2; tuber crop 3; oil 8; vegetable 16; fruit 6
Yangtze rice–wheat	Rice 58; cotton 13; wheat 31; barley 19	Jiangsu	Rice 29; wheat 23; cotton 5; corn 5; bean and pea 5; tuber crop 1; oil 11; vegetable 15; fruit 3
Rice–tea	Rice 73; rapeseed 13	Zhejiang	Rice 35; wheat 2; cotton 0.6; corn 2; bean and pea 7; tuber crop 4; oil 9; vegetable 24; tea 6; fruit 11
Sichuan rice	Rice 41; opium 11; rapeseed 13; wheat 19; corn 14	Sichuan	Rice 22; wheat 13; corn 12; bean and pea 13; tuber crop 14; oil 11; vegetable 12; tea 2; fruit 5
Double-cropping rice	Rice 90; sweet potato 12; sugar cane 6	Guangdong	Rice 44; corn 3; bean and pea 2; tuber crop 8; oil 7; vegetable 24; fruit 21
Southwestern rice	Rice 60; opium 19; corn 14; bean and pea 14	Guizhou	Rice 15; wheat 9; corn 15; bean and pea 7; tuber crop 17; oil 11; tobacco 4; vegetable 10; fruit 2

Source: Data for 1929–1933 are from *Land Utilization in China*, pp. 39–40. Data for 2005 are from *China Statistical Yearbook 2006*, calculated and rearranged by the authors

However, from the perspective of regional characteristics, the major rice production regions, including the middle and lower reaches of Yangtze River, Southeastern China, and Southern China, still dominate rice production, accounting for 60–70% of China's total rice planting area, while the Huanghe-Huaihe-Haihe (HHH) plain dominates production of wheat, accounting for 50% of the total planting area. Northeast China and HHH plain are the major corn production regions, accounting for

40% of the total area respectively, and dominating corn output in the country. Northeast China is also the major soybean production region and has the greatest proportion of soybean planting area. Vegetable production in China enjoys a competitive advantage; in recent years, due to a continuous increase of demand in vegetables caused by a change in diets, the area of planting has continued to increase and now accounts for over 15% of the total planting area of local crops, largely substituting for that of grain crops. Increases in fruit and tea planting areas are closely related to the endowment of local resources. The production of these advantageous crops has created a diversified cropping structure in various regions.

12.3.4 Changes in MCI

MCI is the ratio of sown area to cropland area in a certain region over a certain period (usually one year). It reflects the degree of utilizing agricultural resources in agricultural production and is generally more than one. Substantially, from the angle of cropland utilization, MCI reflects the status and degree of cropland utilization by cropping system. By looking at the MCI of cropland, we can observe if, and to what degree, farm households have altered the intensity of multiple cropping.

Table 12.5 shows that the MCI of China's farm households in 2005 increased by 11% on average, compared with that of 1930s. This suggests,

Table 12.5 Comparison of cropland MCI between 1929–1933 and 2005

Agricultural area (year 1929–1933)	*MCI*	*Area (year 2005)*	*MCI*
Average of eight areas	1.390	National average	1.543
Spring wheat	0.932	Gansu	1.118
Winter wheat–millet	1.153	Shaanxi	1.412
Winter wheat–kaoliang	1.320	Shandong	1.544
Yangtze rice–wheat	1.641	Jiangsu	1.721
Rice–tea	1.705	Zhejiang	1.646
Sichuan rice	1.717	Sichuan and Chongqing	1.903
Double-cropping rice	1.761	Guangdong and Hainan	1.840
Southwestern rice	1.420	Guizhou	1.720

Source: *Land Utilization in China* and *China Rural Statistical Yearbook*, calculated and rearranged by the authors

on the one hand, that improvement in agricultural infrastructure, in particular irrigation mechanization, and technological advances in seed quality, fertilizers, herbicides, and pesticides, have enabled farm households to increase the intensity of cropland utilization. On the other hand, a continuous increase in food demand and a decline in cropland area due to urbanization and industrialization, mean that farm households must continually adapt traditional practices to deal with agricultural problems. Strong technical support of scientific discoveries is a major driver of increased crop diversity and of maintaining grain output at a supply level sufficient to support households, while adapting to changes in demand. However, a high MCI does not necessarily lead to a better outcome since the scale of operations for each more diversified crop mix will be lower. Without appropriate policies and measurements to maintain the agricultural ecological system, too high MCI will exhaust soil fertility and affect the sustainability of land utilization. Therefore, MCI for different regions should be reasonably determined according to the demand of agricultural products and the carrying capacity of the environment.

12.3.5 Change in Intensity of Cropland Utilization

Intensity of cropland utilization is defined as the structure and quantity of various production factors (inputs) applied to a unit cropland area, for example, input of livestock unit,[1] labor, machinery, and capital per mou or hectare. By comparing the intensity of cropland utilization between 1929 and 1933 and the present we can gain some insights into changes in production practices, input use, and production response. In this way, the intensity of cropland utilization is an important index for judging the growth mode of China's agriculture.

As shown in Table 12.6, although animal machinery was used in agricultural production during the 1920s and 1930s, the number of livestock units was low, with less than one unit per farm household. Farm work relied heavily on the input of human laborers and work time. As pointed out by some foreign scholars, the traditional agriculture of China used less labor animals than European agriculture did and could be called "animal-free agriculture." Data from 2005 indicates the substitution of labor animals

[1]According to definitions in *China Rural Economy* (Buck 1930), p. 19, in this chapter, one cow, mule, or horse is one livestock unit, one buffalo represents 2/3 livestock unit, and one donkey represents 1/2 livestock unit.

Table 12.6 Comparison of various agro-production factors input per household between 1929–1933 and 2005

Agricultural area	*1929–1933*		*2005*			
	Livestock unit	*Region*	*Livestock unit*	*Agricultural machinery (kw)*	*Fertilizer (ton)*	*Pesticides (kg)*
Average of eight areas	0.84	National average	0.53	2.56	0.19	5.8
Spring wheat	0.80	Gansu	1.10	2.88	0.16	4.5
Winter wheat–millet	0.75	Shaanxi	0.10	1.86	0.21	1.4
Winter wheat–kaoliang	0.85	Shandong	0.52	4.27	0.23	7.6
Yangtze rice–wheat	0.53	Jiangsu	0.03	1.95	0.21	6.5
Rice–tea	0.61	Zhejiang	0.02	1.70	0.08	5.3
Sichuan rice	1.76	Sichuan	0.56	1.02	0.11	2.8
Double-cropping rice	0.78	Guangdong	0.21	1.20	0.13	5.6
Southwestern rice	1.14	Guizhou	0.31	1.02	0.10	1.2

Source: Data for 1929–1933 are from *Land Utilization in China*, p. 316. Data for 2005 are from *China Rural Statistical Yearbook 2006*, calculated and rearranged by the authors

by agricultural machinery, with a nearly complete substitution effect in economically well-developed regions such as Jiangsu, Zhejiang, and Guangdong, where the total livestock units (TLU) dropped dramatically. This may be due to spillover effects in the sense that capital availability and non-farm wage-earning capabilities might improve the process of agricultural mechanization faster in regions with greater economic growth. Though other regions maintain a certain quantity of livestock, for the most part TLU in these regions are beef cows or beef and labor cows. However, it is difficult to calculate the accurate quantity of labor animals because further differentiation cannot be made on the statistical data.

A comparison of other production factors reveals that modern agricultural production factors are more widely and heavily used now as compared with 1930s, even though the per household cropland area in

2005 was only a third of that in 1929–1933. This provides a better explanation for the extent of increase in input of agricultural production factors to unit area of land.

12.4 Comparison of Cropland Productivity

12.4.1 *Change in Cropland Productivity*

In 1920s–1930s China, agricultural production technology was backward and there were few, if any, inputs based on biological, genetic, or chemical improvements. Meanwhile, with agricultural infrastructure in a very poor conditions, farm households lacked the capability to withstand natural disaster and agricultural production was completely constrained by natural conditions. After the founding of the People's Republic of China (PRC), improvement of infrastructure and advances in agricultural technology led to a great increase in grain yield per hectare. Comparison of yield per hectare for these two periods reveals a surprising increase of cropland productivity (Table 12.7). The yield per hectare in 2005 was four times that of 1929–1933 for wheat and around twice as much for other crops, including rice.

12.4.2 *Change in Agricultural Labor Productivity*

Agricultural labor productivity is usually defined as the output of unit labor of an agricultural laborer and is generally calculated using the quantity of agricultural products produced by unit labor in unit work time (usually one year).

In comparing the agricultural labor productivity of these two periods, the per labor agricultural output for 1929–1933 is approximately measured by the per labor cereal output (Table 12.8). The data for 2005 is taken as the ratio of the total output of corresponding products to the population engaged in agriculture, forestry, animal husbandry, and fishery. Using these measures of per labor cereal output we found an increase of two to three times over this period. As there were very few livestock products and cash crops during the 1920s–1930s, the increase of both of these indicates that the growth in labor productivity is not just related to cereal products. Compared to that of 1929–1933, labor productivity in 2005 increased surprisingly.

Table 12.7 Comparison of yields of main agro-products between 1929–1933 and 2005

Agricultural area	*Rice*		*Wheat*		*Soybean*		*Peanut*		*Cotton*	
	1929–1933	*2005*	*1929–1933*	*2005*	*1929–1933*	*2005*	*1929–1933*	*2005*	*1929–1933*	*2005*
Average	3400	6260	1100	4275	800	1705	1600	3076	600	1129
Spring wheat—Gansu	–	8059	–	2646	600	1594	–	1722	–	1728
Winter wheat–millet—Shaanxi	2700	6064	900	3312	600	763	–	2563	200	1107
Winter wheat–kaoliang—Shandong	2500	7997	1000	5492	1000	2727	1600	4068	600	1000
Yangtze rice–wheat—Jiangsu	3200	7725	1100	4325	1000	2266	2000	3182	700	873
Rice–tea—Zhejiang	2900	6869	700	3249	700	2265	1500	2676	600	1204
Sichuan rice—Sichuan	3700	7213	1500	3386	–	2473	1000	2349	–	888
Double-cropping rice—Guangdong	2300	5225	1000	2040	500	2252	1400	2452	–	–
Southwestern rice—Guizhou	5000	6552	1500	1778	800	1236	1900	1694	–	398

Source: Data for 1929–1933 are from ***Land Utilization in China***, p. 282. Data for 2005 are from ***China Rural Statistical Yearbook 2006***

Note: Data for 1929–1933 are data for each of the agricultural areas, data for 2005 are data of each of the provinces

Table 12.8 Per labor agricultural output in 1929–1933 and 2005 (kg)

Agricultural area	*Per labor cereal output*	*Per labor agricultural output in 2005*						
			Food	*Cotton*	*Oil*	*Meat*	*Aquatic products*	*Milk*
Average of eight areas	595	National average	1615	19	103	205	170	92
Spring wheat	336	Gansu	1097	15	66	99	2	41
Winter wheat–millet	475	Shaanxi	1090	8	47	94	8	118
Winter wheat–kaoliang	617	Shandong	1797	39	167	222	338	86
Yangtze rice–wheat	580	Jiangsu	2498	28	190	213	343	50
Rice–tea	712	Zhejiang	986	3	61	158	585	32
Sichuan rice	710	Sichuan	1357	1	98	220	42	25
Double-cropping rice	547	Guangdong	915	–	51	238	456	8
Southwestern rice	782	Guizhou	894	0	66	120	7	3

Source: Data for 1929–1933 are from *Land Utilization in China*, p. 380. Data for 2005 are from *China Statistical Yearbook 2006*, calculated and rearranged by the authors.

12.5 Conclusion

The farm household operating scale of China in 2005 tended to be a more scattered and smaller scale operation. The per household cropland area in 1929–1933 was about 1.51 hectares, while in 2005, it was only 0.50 hectares, approximately 30% of that observed in 1929–1933. Part of this change was due to population growth and the implementation of the LUR in the post-collective period. The yield per hectare in 2005 was four times that of 1929–1933 for wheat and around twice for other crops including rice, indicating an increase by nearly two to four times in cropland productivity and more than three times in labor productivity.

The improvement of agricultural productivity of China was not accompanied with a scale expansion of agricultural operation. Multiple cropping and advances in biological and chemical technology of seed, chemical fertilizer, and pesticide have been decisive factors for improvement of agricultural productivity. So far, the use of machine technology to expand the scale of operations has not been a key factor in China's agriculture development.

References

Buck, J. Lossing. 1930. Chinese Rural Economy. *American Journal of Agricultural Economics* 12 (3): 440–456.

Buck. 1937a. *Land Utilization in China: A Study of 16,786 farms I 168 localities, and 38,256 farm families in twenty-two provinces in China, 1929–1933.* University of Nanking (1937), University of Chicago Press, Chicago (1937), and The Council on Economic and Cultural Affairs, NY NT, 1956.

———. 1937b. *Land Utilization in China-Atlas.* Chicago: University of Nanking (1937), University of Chicago Press.

———. 1937c. *Land Utilization in China-Statistics.* Chicago: University of Nanking (1937), University of Chicago Press.

CHAPTER 13

A Comparison of Certain Changes in Chinese Agricultural Operations between Buck's Republican Era and Modern China

Hao Hu and Funing Zhong

13.1 Introduction

In this chapter we analyze changes in ***Land Utilization in China*** and the operation status of farm households by comparing Buck's household data to contemporary statistical and survey data on the agriculture of China. Due to a lack of statistical data for the first half of the twentieth century, there have been few opportunities to undertake comparative economic analyses, therefore, it is important to compare Buck's survey data with current statistics.

We undertake our comparative analyses on a regional basis. In his survey of China's land utilization status, Buck divided the sample regions into two major agricultural zones—the wheat zone and rice zone, which were further divided into eight areas, namely, Spring Wheat Area, Winter Wheat–Millet Area, Winter Wheat–Kaoliang Area, Yangtze Rice–Wheat

H. Hu (✉) • F. Zhong
College of Economics and Management, Nanjing Agricultural University, Nanjing, China
e-mail: huhao@njau.edu.cn

H. Hu et al. (eds.), *Chinese Agriculture in the 1930s*,
https://doi.org/10.1007/978-3-030-12688-9_13

Area, Rice–Tea Area, Sichuan Rice Area, Double-Cropping Rice Area, and Southwestern Rice Area (see also Fig. 2.1 and 2.2 in Chap. 2). The division of agricultural zones and areas were based on basic agricultural conditions and crop varieties. As zoning at Buck's era was quite different from current agricultural regionalization, data for exactly the same areas are unavailable. For purpose of vertical comparison of agricultural development, we use Buck's survey data for an area and the present data of a specific province which is a typical one for Buck's area as a comparison. Specifically, Gansu Province is selected as a typical province for the Spring Wheat Area, Shaanxi Province for the Winter Wheat–Millet Area, Shandong Province for the Winter Wheat–Kaoliang Area, *Jiangsu Province* for the Yangtze Rice–Wheat Area, Zhejiang Province for the Rice–Tea Area, Sichuan Province for the Sichuan Rice Area, Guangdong Province for the Double-Cropping Rice Area, and Guizhou Province for the Southwestern Rice Area. On the basis of these comparisons, we draw on survey data of Jiangning County and Wujin County (now renamed as Jiangning District and Wujin District, respectively) of *Jiangsu Province* for the two periods to compare some micro indices of agricultural operations. In Buck's original survey data, the survey of *Jiangsu Province* focused on these two counties.

The indexes selected in this chapter for comparison include essential agricultural production factors such as labor, land, and capital. Investigation of capital is focused on input of agricultural machinery (farming labor animals). On the basis of these indexes, cropping structure, agricultural productivity, farm household income sources, and part-time farm work are compared to analyze the change in China's agriculture operation over nearly a century.

13.2 Change in Input of Agricultural Production Factors

13.2.1 Change in Input of Agricultural Labor

After several decades of economic development, the proportion of rural population in China decreased from 79% to 58%, a smaller decline compared with some other countries. On the one hand, this is attributed to the weak foundation of China's economy, particularly its industrial foundation, which, despite being vast in the modern era, is still insufficient in its ability to transfer the country's population from the agricultural sector to the industrial sector. On the other hand, China's large population relative to land has likely slowed the country's agricultural development, and objectively delayed the transfer of her agricultural population.

Comparing the change in rural households in Jiangning District and Wujin District, we can further observe the alteration in per household labor quantity, family size, and agricultural population. Tables 13.1 and 13.2 present the average family size of farm household and per household labor quantity in the two districts. It is observed that the average family size of farm household has become smaller since the 1929–1933 period.

One of the major reasons is the one-child family planning policy implemented in recent years. The other is the absolute limitation of cropland resources in the post-collective period and increase in the number of farm households under the 1978 household responsibility system, leading to a decline of per household cropland and decrease in total income of farm households. In spite of decreases in family size and, consequently, a decline in households' agricultural labor supply, there was considerable underemployment in agriculture due to sharp increases in the numbers of farm households and the total quantity of agricultural laborers.

Table 13.1 Urban and rural population distribution for the two periods (%)

Percentage of urban population		*Percentage of rural population*	
1929–1933	*2004*	*1929–1933*	*2004*
21	42	79	58

Source: Data for 1929–1933 are from *Land Utilization in China* (John L. Buck), p. 505, sorted and rearranged by the authors. Data for 2005 come from *China Rural Statistics Yearbook*, 2005, p. 93

Table 13.2 Average family size and per household labor quantity of Jiangning District and Wujin District

Localities	*Per household labor quantity (person)*		*Average family size (person)*	
	1921–1923	*2005*	*1921–1923*	*2005*
Jiangning	4.59	1.79	6.13	3.12
Wujin	3.73	1.57	4.87	3.04

Source: Calculations by the authors based on *Chinese Farm Economy* and *Jiangsu Province Rural Statistics Yearbook*, 2006

Table 13.3 Comparison of educational level of rural population between the two periods (%)

Sex	*1929–1933*			*2004*			
	Uneducated	*Little educated*	*Unknown*	*Illiterate*	*Primary school*	*Middle school*	*High school and above*
Male	54.1	45.2	0.7	7.5	29.2	50.4	12.9
Female	97.7	2.2	0.1				

Source: Data for 1929–1933 are from *Land Utilization in China* (John L. Buck), p. 521. Data for 2004 come from *China Rural Statistics Yearbook*, 2005

13.2.2 Change in Agricultural Laborers' Educational Levels

As shown in Table 13.3, the educational level of farmers in China has risen significantly in recent years, compared with that in 1929–1933, when most farmers were illiterate or semi-illiterate. Due to the lower and "humble" status of women in a patriarchal society, almost no women could read, except a very few from rich families and/or who had open parents. Since 1949, following the introduction of nine years of compulsory education in rural areas, the percentage of population with an educational level at, or above, primary middle school has continuously increased, rising to above 60% in 2004. Though the percentage of illiterate or semi-illiterate persons in the agricultural population has decreased sharply, due to a huge population base, the absolute numbers of such persons was still as high as 100 million in 2004, equivalent to half of the uneducated population in the 1920s–1930s. Uneducated, these farmers were unable to transfer to other industrial sectors, and had to stay in rural areas, remaining engaged, in various ways, in agricultural production. Furthermore, their general circumstances meant they could barely provide the necessary human capital required for driving agricultural growth forward.

13.3 Change in Input of Cropland by Farm Household

In *Chinese Farm Economy*, John L. Buck (1930) measured the scale of household operations using farm size, cropland area, and sowing area. In this chapter, we first compare cropland area, and then sowing area, based on his household survey data.

Table 13.4 Comparison of per household cropland area between 1929–1933 and 2004 (hectare)

Item	*1929–1933*	*2004*
Average	1.51	0.52
Spring Wheat Area/Gansu	2.96	1.08
Winter Wheat–Millet Area/Shaanxi	1.50	0.73
Winter Wheat–Kaoliang Area/Shandong	2.06	0.38
Yangtze Rice–Wheat Area/Jiangsu	1.42	0.32
Rice–Tea Area/Zhejiang	0.88	0.17
Sichuan Rice Area/Sichuan	1.27	0.46
Double-Cropping Rice Area/Guangdong	0.88	0.21
Southwestern Rice Area/Guizhou	0.81	0.62

Source: Data for 1929–1933 are from *Land Utilization in China* (John L. Buck), p. 352. Data for 2004 come from *China Rural Statistics Yearbook*, 2005, rearranged by the authors

Note: Due to lack of statistics data, the per household cropland area for 2004 was calculated by the cropland area of each province divided by the total number of households in the province

Table 13.5 Comparison of per household cropland area and crop planting area (hectare)

Locality	*Cropland area*		*Crop planting area*	
	1921–1925	*2005*	*1921–1925*	*2005*
Jiangning	2.12	0.32	3.78	0.49
Wujin	1.14	0.24	2.05	0.24

Source: Calculation by the authors based on *Chinese Farm Economy* and *Jiangsu Province Rural Statistics Yearbook*, 2006

Note: Data for 2005 were obtained from the total area of cropland and total planting area of the two localities divided by the total number of households of the two localities, respectively.

The per household cropland area in 2004 decreased sharply to just a third of that 80 years previously, while remaining larger in the wheat zone than in the rice zone, albeit with a sharper decrease in the rice zone, particularly in economically developed provinces such as Jiangsu, Zhejiang, and Guangdong. This suggests that the contradiction between land and population could become more prominent as the economy develops.

The per household cropland areas and crop planting areas of Jiangning District and Wujin District, compared in Table 13.5 for the two periods, shows a similar trend to that presented in Table 13.4. Both of these localities are located in economically developed southern *Jiangsu Province*, where the cropland area has been reduced more sharply. This shows that per household cropland area decreased sharply due to an increase in the rural population and decrease of total cropland arising from agricultural lands being put to industrial use and urban development. Investigation of the multiple cropping index (MCI) reveals that in the rural areas of the lower reaches of the Yangtze River this was as high as 180% in the 1920s. The MCI in 2004 was around 150% for Jiangning District, which was lower than that of 80 years previously, and even lower for the Wujin District. These changes may be attributed to the change in farmer income sources. In comparison to 1929–1933, farmers in 2004 in this area could obtain significantly more family income from numerous local rural enterprises, instead of relying solely on farming. Consequently, it may not be necessary for them to increase the MCI on limited cropland to improve income.

13.4 Change in Agricultural Capital Input

Input from agricultural machinery (or farming labor animals), the only comparable capital input item in these two periods, is compared to investigate capital input in agricultural production. In order to improve labor productivity and land productivity, farmers have continued to improve agricultural production technology. Since the time of Buck's survey agricultural machinery in China has transformed from labor animal machinery to engine machinery. A comparison of these two types of technologies—both of which substituted for human labor—reveals the impact on agricultural production from mechanization in the two periods.

In the 1920s–1930s, China's agriculture was labor intensive, with some labor animal use (see Chap. 9 for an analysis of labor–land substitution effects). Per household, labor animals are expressed as a measure of total livestock units (TLU), which can reflect the popularity and productivity of such animals (Table 13.6).

Though there was use of labor animals in agricultural production during the 1920s–1930s, it was not extensive, with less than one livestock unit per farm household. As pointed out by foreign scholars, traditional Chinese agriculture used less labor animals than European agriculture did and may be called "animal-free agriculture." In comparison, data from 2004 reveals a strong substitution of labor animals by agricultural machinery, with nearly complete substitution in economically well-developed regions such as Jiangsu, Zhejiang, and Guangdong, where the TLU dropped dramatically. This suggests that agricultural mechanization has been adopted more extensively in regions that experienced faster economic growth. Though other

Table 13.6 Comparison of input of labor animals and agricultural machinery 1929–1933 and 2004 by region

Item	*1929–1933*	*1929–1933*	*2004*
	Per household	*Per household*	*Per household*
	TLU	*TLU*	*agricultural machinery (kW)*
China	0.84	0.53	2.56
Spring Wheat Area/Gansu	0.80	1.10	2.88
Winter Wheat–Millet Area/Shaanxi	0.75	0.10	1.86
Winter Wheat–Kaoliang Area/Shandong	0.85	0.52	4.27
Yangtze Rice–Wheat Area/Jiangsu	0.53	0.03	1.95
Rice–Tea Area/Zhejiang	0.61	0.02	1.70
Sichuan Rice Area/Sichuan	1.76	0.56	1.02
Double-Cropping Rice Area/Guangdong	0.78	0.21	1.20
Southwestern Rice Area/Guizhou	1.14	0.31	1.02

Note: Livestock unit: one cow, mule, or horse is 1 livestock unit, one buffalo represents 2/3 livestock unit, and one donkey represents 1/2 livestock unit (Buck, *Chinese Farm Economy*, p. 19). In Buck's 1929–1933 survey, the animal husbandry production in various areas of China mainly included labor animals and productive animals. The former includes various draft animals, the latter includes animals bred for producing meat, leather, or other products. Buck pointed out that among livestock bred in rural China in that period, the ratio of labor animals to productive animals was approximately 3:1. In other words, most of the livestock bred were labor animals, a finding that is attributed to the productivity and economic development levels of that period

Source: Data for 1929–1933 are from *Land Utilization in China*, p. 316. Data for 2005 come from *China Rural Statistics Yearbook*, 2005, p. 52

regions maintain a certain quantity of livestock, most units in these regions are beef cow or beef and labor cows. However, it is difficult to accurately quantify labor animal productivity as further differentiation from available statistical data is impossible.

13.5 Change in Agricultural Operation

13.5.1 Change in Cropping Structure

The decisive factor of China's agricultural cropping structure in 1929–1933 was the endowment of natural resources from soil and climate. Due to lower levels of labor productivity and unstable social and political environments, the crops planted in that period were mainly rice, wheat, kaoliang, potato, and cotton, which provided food and clothing. Cash crops planted in some regions, for example, opium planted in Southwestern Rice Area and Sichuan Rice Area, were owned by warlords and prominent landlords.

The development of a commodity economy is a decisive factor influencing the cropping structure of modern agriculture. Table 13.7 shows a decline in the proportion of planting areas of staple grain crops such as rice and wheat, as compared with 1929–1933. The planting areas of rice and wheat fell by 15% and 14%, respectively, while the planting areas of corn, soybean, and oil crops saw a substantial increase. However, from the perspective of regional characteristics, the major rice production regions,

Table 13.7 Comparison of percentage of crops' planting area between 1929–1933 and 2004 (%)

Item	*1929–1933*	*2004*
Average	Rice 33; wheat 29; cotton 7	Rice 18; wheat 14; cotton 4; corn 17; bean 8; tuber crop 6; oil plant 9;
Spring Wheat Area/Gansu	Millet 34; spring wheat 18; potato 10	Rice 0.1; wheat 25; cotton 2; corn 13; bean 6; tuber crop 15; oil plant 9
Winter Wheat–Millet Area/Shaanxi	Wheat 40; millet 21; cotton 9	Rice 4; wheat 28; cotton 2; corn 26; bean 9; tuber crop 7; oil plant 7
Winter Wheat–Kaoliang Area/Shandong	Wheat 46; millet 23; cotton 9; kaoliang 19; corn 16	Rice 1; wheat 28; cotton 10; corn 23; bean 2; tuber crop 3; oil plant 8
Yangtze Rice–Wheat Area/Jiangsu	Rice 58; cotton 13; wheat 31; barley 19	Rice 28; wheat 21; cotton 5; corn 5; bean 5; tuber crop 1; oil plant 12
Rice–Tea Area/Zhejiang	Rice 73; rapeseed 13	Rice 37; wheat 2; cotton 0.6; corn 2; bean 7; tuber crop 4; oil plant 9
Sichuan Rice Area/ Sichuan	Rice 41; opium 11; rapeseed 13; wheat 19; corn 14	Rice 22; wheat 13; cotton 0.4; corn 12; bean 13; tuber crop 6; oil plant 12
Double-Cropping Rice Area/Guangdong	Rice 90; sweet potato 12; sugar cane 6	Rice 44; wheat 0.1; corn 3; bean 2; tuber crop 8; oil plant 7
Southwestern Rice Area/ Guizhou	Rice 60; opium 19; corn 14; bean 14	Rice 15; wheat 9; corn 15; bean 7; tuber crop 17; oil plant 11

Source: Data for 1929–1933 are from *Land Utilization in China*, pp. 39–40. Data for 2004 come from *China Rural Statistics Yearbook*, 2005, pp. 458–460

including the middle and lower reaches of Yangtze River, Southeast China, and South China, still dominated, accounting for 60–70% of total rice planting area, while the Huanghe-Huaihe-Haihe (HHH) plain dominated for wheat, accounting for 50% of the total planting area. Northeast China and the HHH plain were the major corn production regions, accounting for 40% of the total corn planting area respectively, and with corn output in these two regions dominating the country as a whole. Northeast China was also the major soybean production region. In addition to these dominant crops, in these major farming areas other cereal crops and cash crops were planted too, forming a more balanced cropping structure.

13.6 Change in Agricultural Productivity Level

13.6.1 Comparison of Labor Productivity

Labor productivity is usually defined as the output per unit labor of a laborer and is generally calculated using the quantity of products produced by unit labor in unit work time. As shown in Table 13.8, the per labor output of cereal crops had increased about twice as much in 2004 compared with that of 80 years previously. There were very few livestock products and cash crops in the 1920s–1930s, so increases in these suggest that the growth of labor productivity was not solely driven by increases in cereal output.

13.6.2 Comparison of Land Productivity

Cereal production in 1929–1933 almost completely relied on natural conditions and was frequently impacted by natural disasters. Moreover, the agricultural production technology was relatively backward in comparison to the modern era as a whole and there was almost no biochemical technology input. Consequently, on the whole, cereal output was quite low (Table 13.9). Since 1949, unit cereal output has increased steadily due to the improvement of agricultural infrastructure and advances in agricultural technology, including improved seeds, fertilizer, mechanization, and so on. A comparison of unit output of crops between the two periods reveal that land productivity has improved significantly. For example, the unit output of wheat and rice in 2004 was four times and twice that in 1929–1933, respectively.

Changes in wheat and rice output in Jiangning and Wujin was compared to investigate the growth of land productivity. The comparison

Table 13.8 Per labor agricultural output in 1929–1933 and 2004 (kg)

Item	*Per labor cereal output in 1929–1933*	*Per labor agricultural output in 2004*					
		Cereal	*Cotton*	*Oil plant*	*Pork, beef, and mutton*	*Aquatic product*	*Milk*
China	595.3	1518	20.4	99.1	186.8	158.5	73.1
Spring Wheat Area/Gansu	336.3	1058	14.4	63.7	92.8	1.9	33.5
Winter Wheat–Millet Area/Shaanxi	475.2	1069	8.5	47.3	92.0	7.6	98.8
Winter Wheat–Kaoliang Area/Shandong	617.1	1582	49.4	166.0	207.4	323.1	72.4
Yangtze Rice–Wheat Area/Jiangsu	580.0	2392	42.5	202.0	205.0	309.6	45.3
Rice–Tea Area/Zhejiang	711.6	982	2.7	57.4	155.0	580.8	30.5
Sichuan Rice Area/Sichuan	710.3	1316	1.4	94.6	220.0	36.0	22.0
Double-Cropping Rice Area/Guangdong	547.4	906	0	50.5	159.5	433.2	7.1
Southwestern Rice Area/Guizhou	782.1	881	0	63.4	104.5	6.8	2.7

Source: Data for 1929–1933 are from ***Land Utilization in China***, p. 380. Data for 2004 come from ***China Rural Statistics Yearbook***, 2005, p. 474

Note: For 1929–1933, only cereal output is available

reveals that the unit output of rice in the two provinces increased by 2.88 and 2.26 times, respectively, and the unit output of wheat increased by 5.28 and 5.2 times respectively. Consistent with Table 13.10, the unit output of wheat increased more than that of rice.

13.7 Change in Commodity Rate of Agricultural Products

It is generally believed that China in the 1920s–1930s had a self-supply type of agriculture, so the commodity rate of agricultural products should have been very low. However, national survey data shows that the commodity rate of agricultural products in that period was 52.6%, lower in the

Table 13.9 Comparison of unit output of main cereal crops between 1929–1933 and 2004 by areas (kg/hectare)

Item	*Rice*		*Wheat*		*Soy bean*		*Peanut*		*Cotton*	
	1929–1933	*2004*	*1929–1933*	*2004*	*1929–1933*	*2004*	*1929–1933*	*2004*	*1929–1933*	*2004*
Average	3400	6310.6	1100	4252	800	1815	1600	3022	600	1111
Spring Wheat Area/Gansu	–	7881.9	–	2917	600	1622	–	1891	–	1609
Winter Wheat–Millet Area/Shaanxi	2700	5967.1	900	3560	600	1052	–	2563	200	1027
Winter Wheat–Kaoliang Area/Shandong	2500	7280.4	1000	5338	1000	2972	1600	3948	600	1036
Yangtze Rice–Wheat Area/Jiangsu	3200	7918.8	1100	4295	1000	2635	2000	3162	700	1228
Rice–Tea Area/Zhejiang	2900	6682.0	700	3198	700	2280	1500	2668	600	1213
Sichuan Rice Area/Sichuan	3700	7363.8	1500	3311	–	2444	1000	2277	–	924
Double-Cropping Rice Area/Guangdong	2300	5250.7	1000	2850	500	2252	1400	2482	–	–
Southwestern Rice Area/ Guizhou	5000*	6657.3	1500	1790	800	1364	1900	1866	–	399

Source: Data for 1929–1933 are from ***Land Utilization in China*** (John L. Buck), p. 282. Data for 2004 comes from ***China Rural Statistics Yearbook***, 2005, pp. 156–159

Note: Original note "*The output of rice is extremely high because land of two districts in this area is very fertile and there might be deviation in conversion of nonstandard local land area units to hectares."

Table 13.10 Change in unit output of rice and wheat in Jiangning District and Wujin District between 1921–1925 and 2004 (kg/hectare)

Locality	*Rice*		*Wheat*	
	1921–1925	*2005*	*1921–1925*	*2005*
Jiangning	2454	7075	744	3930
Wujin	3576	8082	696	3617

Source: Calculation by the authors based on *Chinese Farm Economy* and *Jiangsu Province Rural Statistics Yearbook*, 2006

Table 13.11 Comparison of commodity rates of agricultural products between the two periods (%)

Locality	*1921–1925*	*2005*
Jiangning	70.05	62.74
Wujin	46.3	62.74

Source: Calculation by the authors based on *Chinese Farm Economy*, p. 275, and *Jiangsu Province Rural Statistics Yearbook*, 2006

Note: Due to unavailability, data for the two districts for 2005 were substituted by the mean values of *Jiangsu Province*.

northern regions (43.5%) and higher in the eastern and middle regions. As presented in Table 13.11, it was as high as 70% in Jiangning County and 46.3% in Wujin County. Farmers required greater amounts of cash to buy various production materials and other daily living necessities. This might be the main cause for high commodity rates. Moreover, a surplus of agricultural products provided an opportunity to sell items on the open market.

13.8 Change in Farm Household Incomes

13.8.1 Comparison of Farm Household Income Sources

Table 13.12 presents the income sources of farm households in seven provinces of China. During 1921–1925, the cash income of farm households mostly came from marketing agricultural products, of which cereals made up the most important part. Agriculture was the only source of farmer income, and non-agricultural income was very low. Income, particularly the cash income of farm households in Northern China, was

Table 13.12 Income sources of farm households in 17 counties across seven provinces during 1921–1925 (Yuan)

Localities	*Cereal*	*Fiber crop*	*Cash income*					*Non-cash income*		*Total income*
			Vegetable and rhizome crops	*Fruit*	*Silkworm and mulberry*	*Livestock*	*Miscellaneous*	*Products for family use*	*Increase of capital*	
Mean of northern region	93.7	3.5	7.4	1.9	0.2	8.4	11.3	143.5	8.2	278.3
Mean of eastern and middle region	219.6	7.8	18.7	1.5	14.7	28.5	157.4	6.9	486.5	
Mean of 17 counties	153.0	5.5	12.7	1.7	7.0	17.8	20.8	150.1	7.6	376.2

Note: Unit in this table, as that in Table 13.13, was the currency unit in 1921–1925, when 1 USD = 1.89 Yuan

Source: *Chinese Farm Economy*, p. 84

significantly lower than that of the eastern and middle regions. This difference might be caused by the difference in agricultural products varieties or the higher prices in the eastern and middle regions. For easy comparison with income sources of present farm households, Table 13.13 presents the incomes sources of farm households in Jiangning and Wujin, where cash income accounted for as much as 74.7% and 48.8% of a household's total income, respectively.

Table 13.14 shows the income structure of farm households in Jiangning District and Wujin District for 2005. Non-agricultural income (other income) increased significantly by 66.3% and 53.5% for these two districts, respectively. Agricultural income was no longer the main source of farmers' income.

13.9 Farm Households' Part-Time Non-Farm Work

As shown in the 1929–1933 survey data, farm households also participated in part-time non-farm work (Table 13.15). Buck noted that only a little more than two thirds of the agricultural population were engaged in full-time farm work, one eighth were engaged in sideline work, and one fifth were engaged in both farm work and sideline work. Although there was a great deal of surplus labor, farm households could only undertake non-farm work according to the economic development levels at that time, so the percentage of population engaged in both farming work and non-farm work was higher than those only engaged in farm work. A comparison across regions reveals that because labor intensity during the wheat growth period was lower than that during the rice growth period, the percentage of farm households engaged in part-time non-farm work was higher in the wheat zone than in the rice zone. About 16% of adult men in the wheat zone and 11% of adult men in the rice zone were employed in non-farm jobs.

In post-1978 China, the household responsibility system and allocation of land use rights (LUR) witnessed a decrease in the scale and size of farm households. Consequently, there were not only seasonal labor surpluses but also a great number of absolute labor surpluses. The economic rational of farm households to maximize income provided strong incentives to transfer surplus labor away from agriculture, leading to the rapid development of sideline operations, while meeting the increased labor demand

Table 13.13 Income sources of farm households during 1921–1925 (Yuan)

Localities	*Cereal*	*Fiber crop*	*Cash income*					*Non-cash income*		*Total income*
			Vegetable and rhizome crops	*Fruit*	*Silkworm and mulberry*	*Livestock*	*Miscellaneous*	*Products for family use*	*Increase of capital*	
Jiangning	217.6	0.15	13.6	1.5	41.9	27.1	43.5	115.9	1.2	462.5
Wujin	95.5		1.4	3.1	53.1	30.5	4.6	194.0	3.4	385.6

Source: *Chinese Farm Economy*

Table 13.14 Structure and amount of farm household income in 2005 in Jiangning District and Wujin District (Yuan)

Locality	*Agricultural income*	*Other income*	*Total*
Jiangning	2193	3427	5620
Wujin	3292	3781	7073

Source: Calculation by the authors based on *Jiangsu Province Rural Statistics Yearbook*, 2006

from industrialization and urban development. Data in Table 13.15 shows that in 2004 agricultural income accounted for about 50%–60% of farm households' total income, and below 50% in developed regions such as Zhejiang, Jiangsu, and Guangdong. Along with economic development and increased off-farm employment opportunities, farm households were increasingly engaged in non-farm work.

13.10 Summary

Great changes have occurred in China's agriculture since Buck conducted his national survey on land utilization. By comparing agricultural operations during the Republican era to those in 2004 and, further, comparing changes in agricultural productivity and farm household income, we may draw the following conclusions:

1. The operation scale of China's agriculture has become smaller and more scattered when compared with that of the 1920s–1930s, with per household cropland area and per labor cropland area only about 30% of that in the first period. Though the agricultural labor per household unit decreased in quantity, the absolute increase of the rural population has substantially increased surplus rural labor. This surplus may be in excess of 100 million people. There has also been significant growth in agricultural machinery, but in 2004/2005 animal machinery still existed in many regions. Compared to the adoption of other agricultural technologies (e.g. seed and fertilizer), the advance of agricultural machinery technology appears to have been slower.
2. In post-1978 China, agricultural production levels improved dramatically. This was specifically reflected by the improvement of land productivity by two to three times and labor productivity by over three

Table 13.15 Comparison of percentage of farm households doing non-farm work in 1929–1933 and 2004

Item	*1929–1933 2004 (Yuan/person) Percentage of laborers engaged in the following works*								
	Farm work	*Non-farm work*	*Farm work and non-farm work*	*Percentage of farm work in total work*	*Total net income*	*Wage income*	*Net income of household operation*	*Percentage of agricultural income in total income*	*Percentage of wage income in total income*
Mean	68	12	20	80	2936.4	998.5	1745.8	59.45	34.00
Spring Wheat Area/Gansu	69	9	22	79	1852.2	527.6	1228.9	66.35	28.48
Winter Wheat–Millet Area/Shaanxi	54	23	23	69	1866.5	690.4	1028.3	55.09	36.99
Winter Wheat–Kaoliang Area/Shandong	69	12	19	78	3507.4	1178.3	2147.5	61.23	33.59
Yangtze Rice–Wheat Area/Jiangsu	71	7	22	83	4753.9	2443.4	2018.5	42.46	51.40
Rice–Tea Area/Zhejiang	68	12	20	80	5944.1	2855.8	2533.2	42.62	48.04
Sichuan Rice Area/Sichuan	62	15	23	78	2518.9	829.2	1568.3	62.26	32.92
Double-Cropping Rice Area/Guangdong	81	8	11	87	4365.9	2173.2	1805.7	41.36	49.78
Southwestern Rice Area/Guizhou	75	6	19	88	1721.6	505.2	1115.9	64.82	29.35

Note: Due to differences in statistical calibration, data for 1929–1933 means division of different workload, data for 2004 are incomes obtained from different work

Source: Data for 1929–1933 are from ***Land Utilization in China***, p. 390. Data for 2004 come from ***China Rural Statistics Yearbook***, 2005, p. 361

times. As the development of China's agriculture did not rely on increasing the scale of operations we observe no real economies of scale. However, advances in biotechnology, the development of hybrid seeds, expanded research and development in agricultural areas, and expanded availability of chemical fertilizers and pesticides has been decisive in the improved development of China's agriculture.

CHAPTER 14

Concluding Thoughts

Hao Hu, Funing Zhong, and Calum G. Turvey

14.1 Introduction

In this book we have provided a first-round analysis of the recently discovered and digitized microdata compiled by John Lossing Buck in his 1937 three-volume publication *Land Utilization in China*. In doing so we have provided the reader with a general background to the life and work of Buck, the conditions of conflict and calamities that beset China at the time, and the challenge of preserving the data.

This was followed by a number of analytical chapters, including an assessment gauging the accuracy and representativeness of the data and comparisons between China's agriculture in the 1930s and the present day, with a focus on, among other things, cropping structures and changes in productivity. Issues of tenancy were never far from the fore of the farm economy and it is important to understand the contracting complexities of

H. Hu • F. Zhong
College of Economics and Management, Nanjing Agricultural University, Nanjing, China

C. G. Turvey (✉)
Cornell University, Ithaca, NY, USA
e-mail: cgt6@cornell.edu

H. Hu et al. (eds.), *Chinese Agriculture in the 1930s*,
https://doi.org/10.1007/978-3-030-12688-9_14

landlord–tenant relationships and how these affected living conditions, production decisions, and risk. While tenancy is a common theme in most of the major writings about China's agriculture in the 1930s, so too is the role of labor, surplus labor, and the relationship between human and animal labor. Understanding productivity is also important and in Chap. 10 we investigate the relationship between farm size and crop production, while Chap. 11 looks at the scale efficiencies of the farming operation. Chapters 12 and 13 examine household impacts, including estimates of the Gini and Engel coefficients to measure income inequality within the farm sector and empirical estimates of credit and credit demand. In the remainder of this concluding chapter we summarize the key findings.

14.2 Representativeness of Buck's Data

We would first like to address the issue of the reliability of Buck's data. Raising doubts about data is not difficult. The harder question is what is the relevant standard to which such data should be held? From our point of view, Buck's survey was the largest and most stratified sampling ever conducted in rural China, and within the stratification the random selection of 100 households is likely reliable. Even so, the standard deviation about any average reported contains the truer distribution of outcomes, and any sampling within the statistical range is not necessarily statistically different. So, if Huang and Myers reported data differ from Buck's that doesn't mean that their claim is any more correct or that we might claim that their data is incorrect. The essential point is internal consistency. What we find in Huang and Myers and Fei and other writers of the period is that no matter what the data, the story lines match up and are internally consistent in virtually all respects. In response to criticisms, Chap. 5 by Funing Zhong, Hao Hu, and Qun Su investigates the reliability of Buck's data. The authors make a comparison with a completely different set of data, and this data shows that while measures differ for wheat and rice, there is no particular pattern to under or overreporting as some observers have suggested.

One of the critics is Philip Huang, who based his assessment on Japanese data collected during the occupation. The Japanese data used by Huang is indeed very rich. Although gathered for a different purpose than Buck's data, Huang levels the general criticism that Buck's data was not representative and was "really a sample of the well-to-do" (p. 38). As evidence, Huang notes that realistic farm sizes in Hebei and Shandong were 4.2

mou/capita and 3.7 mou/capita compared to Buck's average size of 7 mou in the Winter Wheat–Kaoliang region. In fact, the land reported by Buck (1937a, Table 15, p. 279) finds the per capita average to be 0.56 acres, which, after multiplying by six to obtain a standard mou, is 3.36—only slightly less. There are other points of comparison between the Japanese data and Buck's that bring doubt into question. For example, Buck (1937a, Table 7, p. 272) shows that very large farms in the Winter Wheat–Kaoliang area were between 120 and 153 mou (20.07 and 25.48 acres), which is certainly in the neighborhood of the range of land owned by "rich" managerial farms reported by Huang (Table 10.3. p. 175), which average 141.2 mou. Other comparisons are difficult because Huang focused largely on larger managerial farms.

Myers (1976), also using the Japanese data, concluded that the results of his study were closer to Buck's than any other up to that time. Even so, Myers goes to significant lengths to address the validity of Buck's data by comparing it to 1930 data compiled by the Statistical Bureau of the Department of Agriculture, Industry, and Commerce and published in *Statistical Monthly*. Unsurprisingly, he found an imperfect match but could not determine which was more accurate. Since the Statistical Bureau surveyed significantly more counties than Buck, it would seem reasonable that this data would be more accurate on average. Differences held even when the same counties were compared. The reality is that such comparisons are very difficult. In 1929 and 1930 there was an extreme drought that affected Hebei and Shandong, and this could account for the lower government yield measures gathered in 1930. As Chap. 3 in this book points out, the number of calamities and conflicts over the 1929–1933 period were so extraordinary that to address productivity differences across the years is—at least to our opinion—futile. Nonetheless, with access to individual household yield data, Chap. 5 in this book shows, for Hebei, an average of 98 jin/mou for wheat, which compares favorably with the 91.1 jin/mou for Hebei reported in Myers (p. 317).[1] That Buck recorded the actual, typical, and best yields suggests that he understood that the yields recorded by survey were captured only at a moment in time and did not mirror typical conditions. Indeed, it appears that given the

[1]As a point of reference Myers (1976) presents wheat for Hebei of 141.2 jin/mou. This appears to be drawn from the most frequent wheat yields, not the actual yields, as reported in Buck (1937a, Table 14, p. 225), which, in the Winter Wheat–Kaoliang region, was 15 bushels/acre, which (using 60 lbs/bu) converts to 136.6 jin/mou.

conditions during the survey period crop yields across his survey area were only about 78% of typical yields (Buck 1937a, Table 13, p. 223).

Having spent nearly 20 years on preserving, checking, digitizing, and in line with the various contributors to this volume, we believe we can put the criticisms of the data to rest. Certainly the economic environment between 1929 and 1934 had many challenges and these no doubt would have been impactful on the data recorded. Buck made no claims that the data collected were any more than a snapshot of conditions at that moment in time. To make one-to-one comparisons with *Chinese Farm Economy*, which preceded it, or the Japanese data used by Huang and Myers that followed it, is a misplaced effort if the goal is to come to a conclusion about reliability or representativeness. The environmental, ecological, economic, political, and geopolitical dynamism of the Republican era precludes any such comparison. Having said this, what have we learnt?

14.3 Summary of Findings

On tenancy Minjie Yu and Hao Hu provide evidence for Northwest China, confirming that tenancy for landless tenants and small part-owner farmers, was a "tenancy for subsistence." Tenancy area was limited and the household mainly relied on family labor to perform farm work while off-farm employment played a very important role in a tenant's survival. In contrast, because of the special requirements of certain crops, large part-owner farmers rented in land to "improve living conditions," relying on hired labor to varying extents and being less involved in off-farm employment. As for the tenancy system, share rent was dominant in Northwest China, with its "popularity" in proportion to the number of tenant farmers. Whether cash rent or share rent was used impacted production responsibilities, negotiation ability, risk management, rent-rate sensitivity, pressure from subsistence, and crop varieties cultivated.

The conclusions here are not very different from those in Philip Huang's work. Huang (1985) identified a variety of farms ranging from small peasant to large managerial farms. He posited that the smaller farms did not pursue profit maximization but larger managerial farms did. To describe conditions in the smaller farms he invoked the notion of "agricultural involution." Involution has no clear-cut definition and its identification is perhaps more of an art than a science, but evidence of its existence is identified by an increased labor intensity even if the deployment of such labor exhibits decreasing, or even negative, marginal productivity.

In comparison, managerial farms focused on more usual profit-maximizing optimization and would employ land, labor, and capital to that end.

Our findings for North China are not inconsistent with agricultural involution. Small farms operated at the margins of subsistence, with familial labor on the farm put to full use. The closest economic argument is that smaller farms optimized according to the average physical and value product while larger farms optimized according to the more traditional marginal physical and value product. These contrasting objectives put smaller farms at an economic disadvantage in terms of efficiency; operating at the border of subsistence, or worse, their constraint set included safety-first nutritional constraints that would become binding in poor harvest years and sometimes in normal years. The managerial farms were unlikely to face binding nutritional constraints and had, therefore, the flexibility to specialize in higher valued cash crops.

Does this argument hold with respect to labor? Hao Hu and Weiwei Zheng estimate surplus agricultural labor in the rice and wheat zones in China, including six farming areas. Based on these estimates of surplus labor they analyzed the main factors influencing labor surplus, finding that rates of surplus agricultural labor in the rice zone and in the wheat zone were 69% and 79%, respectively, even though the average farm size in the wheat areas (28.37 mou) were almost twice those of the rice areas (14.42 mou). So even though the land-to-labor ratio is higher in the wheat areas the demand for labor was much higher in the rice areas. They attribute this to two main factors. The first was land tenancy and the second was the multiple cropping index (MCI). They find that land rental rates in the wheat area (mostly Northern China) and the rice area (mostly Southern China) were substantially different, at 16.76% and 45.10% respectively. Tenancy in the form of share rent or cash rent was significantly more pronounced in the wheat area, as discussed. The tenancy relationships and agronomic conditions favored greater diversity in crops in the rice region (MCI = 1.48) than in the wheat region (MCI = 1.09). Crops grown in the north were mainly kaoliang, corn, wheat, and barley, which were staple grains largely used for home consumption and nutritional balancing. With surplus labor one would expect a larger rate of substitution between human and animal labor, but the animals per household were, on average, about the same, with 1.14 and 1.11 for the wheat and rice zones respectively. The degree of specialization was determined by climate and access to irrigated water, which was available for only 21.34% of farms in the wheat area and 72.70% in the rice area.

Hisatoshi Hoken and Qun Su address the relationship between land productivity and cultivated area. In particular they sought to determine whether there was an inverse relationship between the two. The conventional wisdom amongst agricultural and development economists is that with a low land-to-labor ratios, smaller farms have a greater amount of excess labor that can be more intensively applied to cultivation, resulting in higher land productivity. Using Box–Cox transformations they show that the relationship between land productivity and the cultivated areas of most crops was positive, with slopes for wheat and barley being linear or slightly negative, and slopes for rice, rapeseed, and seed cotton appearing to be slightly positive. They conclude that the relationship between cultivated area and land productivity was not the same among crops, and that the differences between the labor intensity and level of commercialization of each crop were strongly related to the (non-)existence of an inverse relationship.

These are important findings for a number of reasons. The assumption that smaller farms are more productive than larger farms can't be taken as a matter of course. While the transformations did not explicitly take into account land-to-labor or land-to-household ratios, the conclusion does bring into question the marginal productivity of labor for smaller farms.

The issues of land-to-labor ratios and productivity were addressed further by Hao Hu and Minjie Yu. Their analysis differs in several respects to that just given. First, they found that the smallest farms do have an average output—measured in kg/mou aggregated across all crops grown rather than crop by crop—and that this holds in all regions, except the Spring Wheat area in which small and median-sized farms have higher productivity than the smallest farms; in all areas the small and median-sized farms had greater productivity than large farms. However, they also provided Cobb–Douglas land and labor elasticity estimates. Land elasticities were mixed, with the higher elasticities generally residing with the small and median and median size farms. Likewise, labor elasticities—while low and in all but one instance less than 1.0—were mixed with the elasticities for small farms, being (statistically) lowest for Spring Wheat, Yangtze Rice–Wheat, and Sichuan Rice areas; highest in the Winter Wheat–Kaoliang, Rice–Tea, and Double-Cropping Rice areas; and about equal to large farms in the Winter Wheat–Millet area. So generally an increase in labor of 1% increased production by less than 1%, but this held across all farm sizes except medium and large farms in the Spring Wheat Area.

Hu and Yu also provided labor-to-land (not land-to-labor) elasticities directly. If involution is an economic argument then we would expect to see more elasticity in smaller than larger sized farms. This holds true in the Winter Wheat–Kaoliang, Rice–Tea, Double-Cropping Rice, and Southwestern Rice areas. But it cannot generally be assumed to be true in all regions, particularly in the Spring Wheat, Yangtze Rice–Wheat, and Sichuan Rice areas.

Parsing the data on a more granular level, the results also show that smaller farms had a larger MCI, and in seven of eight areas used more fertilizer per mou than larger farms, and more or about the same number of labor animals per mou than large farms in all regions. Where larger farms dominated was in the use of hired labor. If hired labor dominates familial labor for larger farms, but the labor elasticities for smaller farms are higher, then this suggests that the marginal product of hired labor is lower than family or household labor. Since the labor elasticity measure is a relative measure of labor input to output it does not distinguish between average productivity and marginal productivity of labor; thus, involution cannot so easily be supported, but given these measurements it cannot be excluded either.

Other chapters in the book dealt with agricultural credit and inequality. Hao Hu and Zhongwei Yang investigate inequality by computing measures of the Gini and Engel coefficients using income from hired labor as an income proxy (Buck did not record household income). Using non-parametric Gaussian kernel density estimation, results show that there was regional inequality but no polarization in the average income of agricultural hired laborers in the 1930s. The average income of agricultural hired laborers gradually decreased with regional development conditions, with the relationship between intraregional inequality in income and economic development level generally following a positive U-shaped trend. Unbalanced development of East China, Central China, and West China was the key factor leading to great inequality in agricultural hired labor income at the national level.

Chapter 11 provides the first empirical examination of credit supply and demand in the 1930s and breaks down demand in terms of production and consumption, with supply and demand being endogenous. Although borrowing to cover a range of special expenditures such as a dowry, birthday, and birth of son were recorded by Buck, by a significant margin, weddings and funerals were the highest special expenditures. With regard to borrowing as a whole, the supply appeared almost perfectly inelastic, with the interest rate being set according to the amount of loan demanded. When considering "who" is borrowing there is no particular strength in any variable except for productivity and some local and time effects. There is a strong relationship between increased productivity and borrowing, and a negative effect with higher interest rates. On total credit demand (given a loan was made) the strongest demand forces were production, weddings, and funerals. This holds when non-productive or consumption loans are examined in isolation. When production loans are examined there is a positive relationship with production, as one would expect, but also a relationship with special expenditures on weddings or the birth of a son, which suggests that there is some fungibility in production loans. However, there is no evidence of a relationship between production loans and funerals.

14.4 Final Thoughts

In the opening chapter we set out the economic environment facing China in the 1930s, and provided an overview of current thinking about where China is situated in terms of low- and high-level equilibrium traps. Along the way we considered the issue of agricultural involution because it provided an alternative or different approach to understanding the era. Because Buck's (1937) data is not dynamic there is no way of directly testing whether China's agricultural economy rested in a low- or high-level equilibrium trap as described in Nelson and Elvin. In (1972, 1984, 1996) either case the outcome is at a level of subsistence that would be consistent with a poverty trap. One of the key questions in the modern era is whether China was in some sort of equilibrium trap throughout this period. The main characteristic of an equilibrium trap is that gains in output are just equal to, or near equal to, changes in population and in the neighborhood of subsistence. The low land-to-labor ratio across China suggests that the population pressures of the day were taking their toll on labor productivity and output. Maynard and Swen's summary of nutrition in Chap. 8 of *Land Utilization* finds that while the majority of counties had nutrition levels above the 2800 calories/day trigger, around 25% to 31% were below this standard. Since the data were based on averages, the actual number of farm households below the nutritional poverty line would be substantially more. A separate investigation by Zhou et al. (2018) finds that the conditions of the day in terms of the endogenous linkages between agricultural productivity, wages, and caloric intake were consistent with a nutritional poverty trap, and included some evidence of involution. The deployment of capital would also be stagnant in a low-level equilibrium trap. Mechanization during this period was rudimentary, with reliance on farm animals and organic fertilizers. It is true that advances were being made in some areas of agricultural production, including improved seed selection, but these efforts, in addition to related attempts at agricultural reconstruction and cooperative credit, were too late to have any meaningful effect on Buck's survey of the economy. For the most part, China's agricultural economy was in a state of increasing returns to scale, at least in terms of land and labor, so the country as a whole had not yet reached the homogenous steady state that comes with constant returns to scale. In most regions labor productivity was low, with labor production elasticities being less than 1.0, and in some cases zero. In the absence of a comprehensive economic theory about involution, it would not be inconsistent to suggest that involution is highest when the labor elasticity is lowest, and diminishes as the elasticity rises toward constant or increasing returns to scale. If it takes the intensive employment of household labor operating at an average product greater than the marginal product to improve nutrition, then it appears that a state of involution would be coincident with the presence of a low-level equilibrium trap. Although we cannot prove definitively as to whether China's agricultural economy was in a low-level equilibrium trap in the 1929–1933 period all indications suggest it was. Was China also in a high-level equilibrium trap? Elvin's model captures a sequence of low-level equilibrium traps that can come about when capital—human and otherwise—falls below a theoretical potential. When, across generations, this potential is ultimately exploited, and the highest forms of technology are deployed, at some point the changes in output equal the change in population at the level of subsistence. This is the high-level equilibrium trap (in terms of high technology) and it is equivalent to a low-level equilibrium trap (low in terms of poverty). To escape from this trap would require a revolution in technological innovation and a push in private and public capital expenditures. These public expenditures would have to target the non-agricultural industrial base in order to draw labor away from agriculture, reducing the land-to-labor ratio and thereby increasing land and labor productivity and improving per capita output. Again, we cannot prove the existence of a high-/low-level equilibrium trap in Buck's era. But the conditions at the time give rise to its possibility. Most certainly the recognition of a stagnating and impossibly deficient agricultural economy throughout the 1920s led to significant investments in science and technology, such as the Cornell–Nanjing partnership, while the push for agricultural reconstruction and credit by the China International Famine Relief Commission and the nationalist government suggest that China was at the tipping point of a poverty trap.

14.4.1 Modern-Day Comparisons

Finally, we opened Chap. 1 with the recognition that to understand China's agricultural economy in the present day, we must also have some understanding of her agricultural history. In Chaps. 12 and 13 we provided a comparison of certain factors: input levels, agricultural productivity, agricultural operation status, and income sources of farm households of China between Buck's 1929–1933 survey period and equivalent measures in 2004. We show that when compared with that of the 1930s, the per household cropland area has decreased to about 30%, but land productivity improved two to three times and agricultural labor productivity improved by more than three times. Farm size changes are due to land reforms in the post-Republican era, including the post-collective household responsibility system. Productivity increases can be attributed to advances in agricultural technology, in particular, the application of mechanical technology and, more significantly, the application of biochemical technology. In addition, the income sources of farm households have become more diversified.

In Chap. 12, Hao Hu and Feng Zhang analyzed the change of China's cropland utilization and productivity over the past century based on comparison of farm household data for the two periods. Their analysis mainly focused on changes in the average amount of cropland per farm household, MCI cropping structure, and degree of intensity. As for land productivity, the analysis dealt mainly with changes in cropland productivity and agro-labor productivity. Compared with 1930s agriculture, the average per household cropland areas today has decreased to 30%, but the land productivity and labor productivity has improved two to three times and over three times respectively. These changes can be attributed to technological progress and the extensive application of chemical fertilizers. If there are lessons to be learned from comparative agricultural histories it is that to reduce rural poverty there must be a balance between technological gains, capital improvements, and an efficient balancing of land, labor, and capital, including credit. Our investigation of Buck's discovered microdata as presented in the various chapters in this book has given credence to these important economic insights.

References

Buck, J.L. 1937a. *Land Utilization in China: A Study of 16,786 farms I 168 localities, and 38,256 farm families in twenty-two provinces in China, 1929–1933.* University of Nanking (1937), University of Chicago Press, Chicago (1937), and The Council on Economic and Cultural Affairs, NY NT, 1956.

———. 1937b. *Land Utilization in China-Atlas.* Chicago: University of Nanking (1937), University of Chicago Press.

———. 1937c. *Land Utilization in China-Statistics.* Chicago: University of Nanking (1937), University of Chicago Press.

Elvin, M. 1972. The High-Level Equilibrium Trap: The Causes of the Decline of Inventions in the Traditional Chinese Textile Industries. In *Economic Organization in Chinese Society*, ed. W.E. Willmott. Stanford, CA: Stanford University Press.

———. 1984. Why China Failed to Create an Endogenous Industrial Capitalism. *Theory and Society* 13 (3): 379–391.

———. 1996. *Another History: Essays on China from a European Perspective.* Broadway, NSW: The University of Sydney.

Huang, P. 1985. *The Peasant Economy and Social Change in North China.* Stanford CA: Stanford University Press.

Myers, Ramon. 1976. Socioeconomic Change in Villages of Manchuria During the Ch'ing and Republican Periods: Some Preliminary Findings. *Modern Asian Studies* 10 (4): 591–620.

Zhou, Li, Jie Sun, and Calum G. Turvey. 2018. *Conflicts, Calamities and Nutritional Poverty Traps in a Peasant Economy: Evidence from Rural China, 1929–1933.* Unpublished mimeo, Nanjing Agricultural University.

Index[1]

[1] Note: Page numbers followed by 'n' refer to notes.

H. Hu et al. (eds.), *Chinese Agriculture in the 1930s*,
https://doi.org/10.1007/978-3-030-12688-9

C

The manufacturer's authorised representative in the EU is Springer Nature Customer Service Centre GmbH, Europaplatz 3, 69115 Heidelberg, Germany. If you have any concerns regarding our products, please contact ProductSafety@springernature.com

Printed and bound by CPI Group (UK) Ltd, Croydon, CR0 4YY
30/09/2026
02225216-0014